Benchmark Papers in Systematic and Evolutionary Biology

Series Editor: Carl Jay Bajema
Grand Valley State Colleges

Volume

1 MULTIVARIATE STATISTICAL METHODS: Among-Groups Covariation / *William R. Atchley and Edwin H. Bryant*
2 MULTIVARIATE STATISTICAL METHODS: Within-Groups Covariation / *Edwin H. Bryant and William R. Atchley*
3 CONCEPTS OF SPECIES / *C. N. Slobodchikoff*
4 ARTIFICIAL SELECTION AND THE DEVELOPMENT OF EVOLUTIONARY THEORY / *Carl Jay Bajema*
5 NATURAL SELECTION THEORY: From the Speculations of the Greeks to the Quantitative Measurements of the Biometricians / *Carl Jay Bajema*

RELATED TITLES IN BENCHMARK PAPERS IN GENETICS

Volume

1 GENETICS AND SOCIAL STRUCTURE: Mathematical Structuralism in Population Genetics and Social Theory / *Paul Ballonoff*
2 GENES AND PROTEINS / *Robert P. Wagner*
3 DEMOGRAPHIC GENETICS / *Kenneth M. Weiss and Paul A. Ballonoff*
4 MUTAGENESIS / *John W. Drake and Robert E. Koch*
5 EUGENICS: Then and Now / *Carl Jay Bajema*
6 CYTOGENETICS / *Ronald L. Phillips and Charles R. Burnham*
7 STOCHASTIC MODELS IN POPULATION GENETICS / *Wen-Hsiung Li*
8 EVOLUTIONARY GENETICS / *D. L. Jameson*
9 GENETICS OF SPECIATION / *D. L. Jameson*
10 HUMAN GENETICS / *William J. Schull and Ranajit Chakraborty*
11 HYBRIDIZATION: An Evolutionary Perspective / *Donald A. Levin*

Benchmark Papers
in Systematic and
Evolutionary Biology / 5

A BENCHMARK® Books Series

NATURAL SELECTION THEORY
From the Speculations of the Greeks to the Quantitative Measurements of the Biometricians

Edited by

CARL JAY BAJEMA
Grand Valley State Colleges

Hutchinson Ross Publishing Company

Stroudsburg, Pennsylvania

Benchmark Papers in Systematic and Evolutionary Biology, Volume 5
Library of Congress Catalog Card Number: 82-15633
ISBN: 0-87933-412-6

85 84 83 1 2 3 4 5
Manufactured in the United States of America.

LIBRARY OF CONGRESS CATALOGING IN PUBLICATION DATA
Main entry under title:
Natural selection theory.
(Benchmark papers in systematic and evolutionary biology; 5)
Includes bibliographies and indexes.
1. Natural selection—Addresses, essays, lectures.
2. Evolution—Addresses, essays, lectures. I. Bajema, Carl Jay, 1937- II. Series.
QH375.N37 575 82-15633
ISBN 0-87933-412-6

Distributed worldwide by Van Nostrand Reinhold Company Inc.,
135 W. 50th Street, New York, NY 10020

CONTENTS

PART III: THE FRENCH TRANSFORMISTS

PART IV: BRITISH THOUGHT ON NATURAL SELECTION THEORY PRIOR TO CHARLES DARWIN

PART V: THE DEVELOPMENT OF THE DARWIN-WALLACE THEORY OF EVOLUTION BY SELECTION: 1831-1858

SERIES EDITOR'S FOREWORD

The volumes of the *Benchmark Papers in Systematic and Evolutionary Biology* series reprint classic scientific papers on the evolution and systematics of organisms. These Benchmark volumes do more than provide scholars with facsimile reproductions or English translations of classic papers on a particular topic in a single volume. The interpretative commentaries and extensive bibliographies prepared by each volume editor provide busy scholars with a review of the primary and secondary literature of the field from a historical perspective and a summary of the current state of the art.

The theories in which the environment plays a crucial role in giving direction to evolution can be divided into three categories: Geoffroyism—the production of heritable adaptive variations by the environment acting directly on the organism; Lamarckism—the production of heritable adaptive variations as the result of the increasing use or disuse of organs by an organism as it experiences new needs or the alteration of existing needs due to changes in its environment; and neo-Darwinism—the production of both harmful and beneficial heritable variations in organisms by causes unrelated to the specific adaptive problems the organisms are encountering followed by the selective reproduction of heritable variations brought about by the ecological interactions each organism has with its environment. Both Lamarckism and Geoffroyism have been discredited by modern scientific knowledge concerning the production of gene mutations and their transmission to offspring. Only the Darwinian theory of adaptive evolution by natural, sexual, and artificial selection has been demonstrated to be consistent with the modern scientific understanding of genetics.

The history of the theory of natural selection is traced in the present volume from the speculations of the ancient Greeks to the last decade of the nineteenth century when the biometricians developed statistical methods, which they used to directly measure the direction and intensity of ongoing selection in nature. This volume, the fifth in the series, is the second of several volumes on selection theory being edited by C. Bajema. *Benchmark Papers in Systematic and Evolutionary Biology* volumes are also being planned for topics such as biogeography, comparative anatomy, comparative embryology, interspecific competition, Lamarckism, macroevolution, mimicry, mutationism, origin of the genetic code, orthogenesis, paleobotany, speciation, and theories of systematics.

CARL JAY BAJEMA

PREFACE

Charles Darwin's theory of adaptive evolution by natural, sexual, and artificial selection is the most important and yet the most widely misunderstood principle in biology. The Benchmark volumes on selection theory that I am editing are designed to help students and scholars gain a better understanding of, first, the history of the controversy surrounding the relative importance of selection as an ecological process generating adaptive evolution and, second, the current state of our scientific understanding of the role that selection plays in giving direction to evolution.

The theory that adaptive evolution occurs as the result of selection operating on heritable variations was proposed by Charles Robert Darwin and Alfred Russel Wallace in 1858. *On the Origin of Species by Means of Natural Selection, or the Preservation of Favoured Races in the Struggle for Life,* which Charles Darwin published the following year, triggered a scientific revolution with respect to our understanding of the origin and continuing evolution of species.

A companion Benchmark volume, *Artificial Selection and the Development of Evolutionary Theory* (Benchmark Papers in Systematic and Evolutionary Biology, vol. 4), describes how Charles Darwin, by reading the literature on the breeding of domesticated plants and animals, was led to the idea that selection could bring about evolution. It was only after Darwin had become familiar with selection as a process that could cause "descent with modification" (evolution) in domesticated species that he realized how powerful selection operating in nature was. He became convinced that selection was powerful enough to bring about evolution when he read *An Essay on the Principle of Population . . .* by Malthus (1826, excerpt in Paper 15). Darwin concluded that the power of selection to bring about evolution in nature resided in competition, the "struggle for existence" produced by the overproduction of offspring.

This volume traces the development of the theory of natural selection from the speculations of the ancient Greeks through the expansion of the theory to explain evolution by Charles Darwin and Alfred Wallace to the last decade of the nineteenth century when the biometricians developed statistical methods that they used to directly measure the direction and intensity of ongoing selection in nature. It is impossible to publish all the important papers on natural selection theory in one manageable volume. A companion volume, *The Theory and Measurement of Natural Selection in the Twentieth Century,* is in preparation, as is one on sexual selection theory and the development of evolutionary theory that covers the same period as the present volume. Volumes on sexual selection and the development of sociobiology, domestication of plants and animals by selection, human evolution by selection, and cultural evolution by selection are also in preparation.

I wish to thank the people who provided assistance while I selected papers for inclusion in this volume and wrote the editor's commentaries on the history of natural selection theory: Peter Gautrey, University Library, Cambridge; David Kohn, Harvard University; Will Provine, Cornell University; Michael Ruse, University of Guelph; and the librarians at Grand Valley State Colleges and elsewhere who located and obtained copies of numerous scholarly publications for me through interlibrary loans.

CARL JAY BAJEMA

CONTENTS BY AUTHOR

NATURAL SELECTION THEORY

Part I

THE SPECULATIONS OF THE ANCIENT GREEKS

Editor's Comments on Papers 1 and 2

1 **ARISTOTLE**
Excerpt from *The Physics*

2 **LUCRETIUS**
Excerpt from *De Rerum Natura, Libri Sex*

What natural causes might account for the origin of species? Philosophers living in ancient Greece speculated about the answer to this question (Osborn, 1894; Zirkle, 1941, 1946; Guthrie, 1957; Classen, 1977). Empedocles (495–435 B.C.) proposed the theory that each contemporary species was the descendant of organisms originally produced by spontaneous generation. He contended that individual organisms were spontaneously generated by the random combining of body organs that were already present in the ancient environment. This spontaneous generation of organs and subsequent spontaneous generation of new species was supposed to have occurred only during a short period in the remote past. Necessity (selection) then eliminated those individuals whose random combination of body organs did not allow them to survive and reproduce (Osborn, 1894). Contemporary species were theorized to be the descendants of the subset of organisms produced by spontaneous generation that was not eliminated by necessity. There is no evidence that Empedocles discussed the possibility that a species could gradually evolve and undergo speciation, that is, could evolve into two or more species.

The centaurs, chimeras, and other creatures of Greek mythology may have served as the inspiration for Empedocles's speculation that all species were spontaneously generated some time in the past by the random combining of body organs (Osborn, 1894, p. 38). The facts that there were no living descendants of these mythological Greek creatures, and that humans, domesticated animals, and wild species occasionally gave birth to "monsters" that could not survive and reproduce probably served as the inspiration for the selective elimination of species component of Empedocles's hypothesis.

Empedocles's theory is a classic example of what the philosopher Karl Popper (1959) calls an untestable and thus a nonscientific theory because the particular causes that spontaneously generated new species only operated during a brief period of time in the distant past. Consequently there is no way that the causes postulated by the theory can be scientifically tested.

The few fragments of Empedocles's great poem that have been found do not discuss his theory of spontaneous generation of organisms-selective elimination (Leonard, 1907; Zirkle, 1941, pp. 74–75; Lambridis, 1976). However, the nature of the missing passage that contained Empedocles's theory concerning the role that selection played in eliminating those species whose members were unfit (i.e., whose random combination of body organs did not allow them to survive and reproduce) can be inferred from extant passages found in the writings of Aristotle and Lucretius. In Book II of *The Physics* Aristotle (384–322 B.C.) referred to the "man-faced oxen" discussed by Empedocles, summarized the Greek case for natural selection, and then attempted to refute that case (Paper 1). In *De rerum Natura,* T. Lucretius Carus (94–50 B.C.), a Roman poet and champion of Epicurean philosophy, hypothesized how maladapted monsters might have been formed by the chance combination of body organs and how they would have perished, leaving only those who were fit enough to reproduce as well as to survive (Paper 2).

Lucretius strongly rejected Lamarckian evolution—that is, adaptive change as the result of the inheritance of acquired characteristics—while Aristotle accepted the idea that acquired characteristics could, on occasion, be inherited (Zirkle, 1946, p. 94). None of the Greek thinkers proposed a comprehensive theory of evolution based on the inheritance of characteristics acquired as the result of the direct effect of the environment, the use and disuse of organs, or "willing" (Zirkle, 1946).

It is interesting to note that Charles Darwin cited Aristotle in his "An Historical Sketch of the Recent Progress of Opinion on the Origin of Species" that he appended to the third and later editions of the *Origin of Species* (1866–1872). Darwin thought that Aristotle was advocating selection rather than opposing it and criticized him by contending that:

> We here see the principle of natural selection shadowed forth, but how little Aristotle fully comprehended the principle is shown by his remarks on teeth . . . as his opinions fluctuated greatly at different periods, and as he does not enter on the causes or means of the transformation of species, I need not here enter on details. (Darwin, quoted in Peckham, 1959, pp. 59–60)

The Greek poet Theognis (6th–5th century B.C.) and the Greek philosopher Plato (428–348 B.C.) both expressed their concern over the hereditary quality of human beings by drawing an analogy between the careful selective breeding of domesticated species by humans and the way that mating and reproduction were occurring in human populations (Roper, 1913; Montagu, 1947). Conscious human choice was involved in both of the two selective processes that Theognis and Plato compared. None of the extant writings of the Greeks contain statements about a possible analogy between the changes in domesticated species brought about by conscious human selection and processes operating in the rest of nature that could bring about selective change in wild species.

The Greek idea that came to dominate philosophical thought about the nature of species was Plato's belief in the existence of unchanging and eternal realities—ideal types *(eidos)*—that were supposed to exist independently of the changing things of the world perceived by human senses. The widespread adoption of Plato's typological or essentialist view of nature by philosophers made it difficult, if not impossible, for them to consider seriously the possibility that the origin of species might be explained scientifically in terms of populations changing as the result of the selective multiplication of individual variations (Mayr, 1959, pp. 2–4; 1972).

REFERENCES

Classen, C. J., 1977, Anaximander and Anaximenes: The Earliest Greek Theories of Change? *Phronesis* **22:**89–102.

Darwin, C. R., 1866–1872, *On the Origin of Species by Means of Natural Selection, Or the Preservation of Favoured Races in the Struggle for Life,* 3rd–6th eds., Murray, London. (Reprinted in M. Peckham, ed., 1959, *The Origin of Species by Charles Darwin: A Variorum Text,* University of Pennsylvania Press, Philadelphia.)

Guthrie, W. K., 1957, *In the Beginning: Some Greek Views on the Origins of Life and the Early State of Man,* Cornell University Press, Ithaca, 151p.

Lambridis, H., 1976, *Empedocles: A Philosophical Investigation,* University of Alabama Press, University, Ala., 192p.

Leonard, W. E., 1907, The Fragments of Empedocles, *Monist* **17:**451–474.

Mayr, E., 1959, Darwin and the Evolutionary Theory in Biology, in *Evolution and Anthropology: A Centennial Appraisal,* B. J. Meggars, ed., Anthropological Society of Washington, Washington, D.C., pp. 1–10.

Mayr, E., 1972, The Nature of the Darwinian Revolution, *Science* **176:**981–989.

Montagu, M. F. A., 1947, Theognis, Darwin and Social Selection, *Isis* **37:**24–26.

Osborn, H. F., 1894, *From the Greeks to Darwin: An Outline of the Development of the Evolution Idea,* Macmillan, New York, 259p.

Peckham, M., ed., 1959, *The Origin of Species by Charles Darwin: A Variorum Text,* University Of Pennsylvania Press, Philadelphia, 816p.

Popper, K. R., 1959, *The Logic of Scientific Discovery,* Basic Books, New York, 480p.

Roper, A. G., 1913, *Ancient Eugenics,* Oxford, 76p.

Zirkle, C., 1941, Natural Selection Before the "Origin of Species," *Am. Philos. Soc. Proc.* **84:**71–123.

Zirkle, C., 1946, The Early History of the Idea of the Inheritance of Acquired Characters and of Pangenesis, *Am. Philos. Soc. Trans.* **35:**91–151.

1

Reprinted from pages 168–187 of *The Physics,* vol. 1, P. H. Wicksteed and F. M. Cornford, trans., William Heinemann Ltd, London and G. P. Putnam's Sons, New York, 1929

THE PHYSICS

Aristotle

CHAPTER VIII

ARGUMENT

[In support of final causality in Nature, Aristotle reviews and criticizes the tacit assumptions and explicit arguments of those who deny its existence; in particular, Empedocles' theory of the origin of animal species. (Empedocles held (1) *that in the period when a world is formed by the action of Love, gaining upon Strife, the parts of animals arise separately and are united by Love in all sorts of casual ways. Monsters like man-headed oxen, ox-headed men, etc., perish; those forms survive which happen to be fitted to survive.* (2) *In the opposite period, when Strife is gaining on Love (as in our world), animal species are developed from undifferentiated* (' *whole-natured* ') *forms produced by fire rising from the earth ' to meet its like ' and mixing with water).*—C.]

[*Editor's Note:* The Greek text appearing on facing pages in the original has been omitted.]

We must now consider why Nature is to be ranked among causes that are final, that is to say purposeful; and further we must consider what is meant by ' necessity ' when we are speaking of Nature. For thinkers are for ever referring things to necessity as a cause, and explaining that, since hot and cold and so forth are what they are, this or that exists or comes into being ' of necessity '; for even if one or another of them alleges some other cause, such as ' Sympathy and Antipathy ' or ' Mind,' he straight away drops it again, after a mere acknowledgement.

So here the question rises whether we have any reason to regard Nature as making for any goal at all, or as seeking any one thing as preferable to any other. Why not say, it is asked, that Nature acts as Zeus drops the rain, not to make the corn grow, but of necessity (for the rising vapour must needs be condensed into water by the cold, and must then descend, and incidentally, when this happens, the corn grows), just as, when a man loses his corn on the threshing-floor, it did not rain on purpose to destroy the crop, but the result was merely incidental to the raining ?[a] So why should it not be the same with natural organs like the teeth ? Why should it not be a coincidence that the front teeth come up with an edge, suited to dividing the food, and the back ones flat and good for grinding it, without there being any design in the matter ?[b] And so with all other organs that seem to embody a purpose. In cases where a coincidence brought about such a combination as might have been arranged on purpose, the creatures, it is urged, having been suitably formed by the operation of chance, survived ; otherwise they perished, and still perish, as Empedocles says of his ' man-faced oxen.'[c]

Such and suchlike are the arguments which may be urged in raising this problem ; but it is impossible that this should really be the way of it. For all these phenomena and all natural things are either constant or normal, and this is contrary to the very meaning of luck or chance. No one assigns it to chance or to a remarkable coincidence if there is abundant rain in the winter, though he would if there were in the dog-days ; and the other way about, if there were parching heat. Accordingly, if the only choice is to assign these occurrences either to coincidence or to purpose, and if in these cases chance coincidence is out of the question, then it must be pur-

[a] [*i.e.*, if we attribute the bad result (destruction of the corn) to accident or necessity, why attribute the good result (growth of the crop) to benevolence ?—C.]

[b] [*Cf. De part. anim.* 661 b 7, Xen. *Mem.* i. 4, where the shape of the teeth is alleged among other evidences of design. W. Theiler, *Zur Geschichte der teleologischen Naturbetrachtung* (Zürich u. Leipzig, 1925), p. 25, argues that both Xenophon and Aristotle borrowed from Diogenes of Apollonia. —C.]

[c] [*Frag.* 61.—C.]

pose. But, as our opponents themselves would admit, these occurrences are all natural. There is purpose, then, in what is, and in what happens, in Nature.

Further, in any operation of human art, where there is an end to be achieved, the earlier and successive stages of the operation are performed for the purpose of realizing that end. Now, when a thing is produced by Nature, the earlier stages in every case lead up to the final development in the same way as in the operation of art, and *vice versa*, provided that no impediment balks the process. The operation is directed by a purpose; we may, therefore, infer that the natural process was guided by a purpose to the end that is realized. Thus, if a house were a natural product, the process would pass through the same stages that it in fact passes through when it is produced by art; and if natural products could also be produced by art, they would move along the same line that the natural process actually takes. We may therefore say that the earlier stages are for the purpose of leading to the later. Indeed, as a general proposition, the arts either, on the basis of Nature, carry things further than Nature can, or they imitate Nature. If, then, artificial processes are purposeful, so are natural processes too; for the relation of antecedent to consequent is identical in art and in Nature.

This principle comes out most clearly when we consider the other animals. For their doings are not the outcome of art (design) or of previous research or deliberation; so that some raise the question whether the works of spiders and ants and so on should be attributed to intelligence or to some similar faculty.[a] And then, descending step by

[a] [At *Met.* 980 b 22, bees and other such species are called 'intelligent' (φρόνιμα) but unteachable, as having no hearing. *De part. an.* 648 a 5, Bees, etc., are more intelligent than some sanguineous animals, because of the thin and cold nature of the fluid corresponding to blood in them. 'Some similar faculty,' perhaps the unreasoning intuition of instinct, which resembles the human faculty of intuition (νόησις).—C.]

step, we find that plants too produce organs subservient to their perfect development—leaves, for instance, to shelter the fruit. Hence, if it is by nature and also for a purpose that the swallow makes her nest and the spider his web, and that plants make leaves for the sake of the fruit and strike down (and not up) with their roots in order to get their nourishment, it is clear that causality of the kind we have described is at work in things that come about or exist in the course of Nature.

Also, since the term ' nature ' is applied both to material and to form, and since it is the latter that constitutes the goal, and all else is for the sake of that goal, it follows that the form is the final cause.

Now there are failures even in the arts (for writers make mistakes in writing and physicians administer the wrong dose); so that analogous failures in Nature may evidently be anticipated as possible. Thus, if in art there are cases in which the correct procedure serves a purpose, and attempts that fail are aimed at a purpose but miss it, we may take it to be the same in Nature, and monstrosities will be like failures of purpose in Nature. So if, in the primal combinations, such ' ox-creatures ' as could not reach an equilibrium and goal, should appear, it would be by the miscarriage of some principle, as monstrous births are actually produced now by abortive developments of sperm. Besides, the sperm must precede the formation of the animal, and Empedocles' ' primal all-generative '[a] is no other than such sperm.

[a] [*Frag.* 62 οὐλοφυεῖς μὲν πρῶτα τύποι χθονὸς ἐξανέτελλον, ' whole-natured forms first arose from the earth ' (Burnet), *rohgeballte Erdklumpen* (Diels); primitive forms of life, in which sex and species were not yet differentiated, composed of water and fire.—C.]

In plants, too, though they are less elaborately articulated, there are manifest indications of purpose. Are we to suppose, then, that as there were 'ox-creatures man-faced' so also there were 'vine-growths olive-bearing'? Incongruous as such a thing seems, it ought to follow if we accept the principle in the case of animals. Moreover, it ought still to be a matter of chance what comes up when you sow this seed or that.

In general,[a] the theory does away with the whole order of Nature, and indeed with Nature's self. For natural things are exactly those which do move continuously, in virtue of a principle inherent in themselves, towards a determined goal; and the final development which results from any one such principle is not identical for any two species, nor yet is it any random result; but in each there is always a tendency towards an identical result, if nothing interferes with the process.[b] A desirable result and the means to it may also be produced by chance, as for instance we say it was 'by luck' that the stranger came and ransomed the prisoner before he left,[c] where the ransoming is done as if the man had come for that purpose, though in fact he did not. In this case the desirable result is incidental; for, as we have explained, chance is an incidental cause. But when the desirable result is effected invariably or normally,

[a] [In this final argument Aristotle brings together all three kinds of 'causation': Luck, Nature, Art. By Luck a desirable result which serves a purpose is achieved incidentally and once in a way. In Nature, since desirable results are produced normally or invariably, *a fortiori* they must be purpose-serving, though we cannot see the purposing agency. Even in Art we can imagine the purpose-serving process carrying itself out without the previous conscious design which is certainly present.—C.]

[b] [ἀφ' ἑκάστης δὲ κτλ. might possibly mean: 'The final development reached from any one principle (*e.g.*, human seed) is neither (exactly) the same for every individual (for no two men are exactly alike), nor yet is it any random result (*e.g.* dog or horse); there is, however, in each species, always a tendency towards an identical result (*e.g.* the perfect human type that Nature strives, but never successfully, to produce), if nothing interferes.'—C.]

[c] [W. translated λουσάμενος, but λυσάμενος is the better reading. The allusion cannot be to the Μισούμενος of Menander (*Oxyr. Pap.* No. 1013), to which Simplic. 384. 14 refers. Philoponus 324. 21 suggests the ransoming of Plato at Aegina (Diels compares Diog. L. iii. 20 λυτροῦται δὴ αὐτὸν κατὰ τύχην παρὼν 'Αννίκερις ὁ Κυρηναῖος).—C.]

it is not an incidental or chance occurrence; and in the course of Nature the result always is achieved either invariably or normally, if nothing hinders. It is absurd to suppose that there is no purpose because in Nature we can never detect the moving power in the act of deliberation. Art, in fact, does not deliberate either,[a] and if the shipbuilding art were incorporate in the timber, it would proceed by nature in the same way in which it now proceeds by art. If purpose, then, is inherent in art, so is it in Nature also. The best illustration is the case of a man being his own physician, for Nature is like that—agent and patient at once.

That Nature is a cause, then, and a goal-directed cause, is above dispute.

CHAPTER IX

ARGUMENT

[*Most philosophers recognize in the process of natural generation only an 'unconditional,' mechanical necessity, whereby the forces inherent in matter blindly produce their effects. Aristotle elsewhere (e.g.* De part. anim. 642 a 1) *admits that many things are produced by the 'simple' necessity of material and efficient causes alone; but he argues that we must also recognize a 'hypothetical' necessity, seen in the processes of Nature, as in those of art, where purpose fixes a goal to which the mechanical chain of causation shall lead, rather than in other directions.* If *a given desirable result is to be achieved, such and such matter is necessary. The existence of the matter and its forces is a condition* sine qua non *of the realization of such ends, but will not account for the process of realization, which is explained and directed by the end to be brought about. (Cf. the fuller discussion in* De part. anim. i. 1).

The physicist is primarily concerned with the form, definable essence, or end; but where the form 'necessitates' a certain kind of matter, his definitions will include mention of the appropriate matter.—C.]

The phrase 'must of necessity' may be used of what is unconditionally necessary or of what is 'necessary to this or that.'[b] Of which kind is the

[a] [*Cf. Eth. Nic.* iii. 3. In some arts (*e.g.* writing) there is no deliberation; the right spelling of a word is not debatable. In others (*e.g.* medicine) we deliberate about means (how to heal), not about the end (healing).—C.]

[b] [The 'hypothetically' necessary is described as 'the *sine qua non* of a desirable result' (οὗ οὐκ ἄνευ τὸ εὖ, *Met.* 1072 b 12)—necessary, *if* that result is to be achieved.—C.]

necessity that exists in Nature? For people ignore this distinction and talk about things being 'necessarily generated,' much as if they thought that a wall would come up in the necessary course of things because what is heavy naturally descends and what is light is naturally on the top, so that stones go down and make the foundations, while the lighter brick [a] rises above them, and the timber, lightest of all, roofs them above.

No doubt it is a fact that the building cannot dispense with these materials, and in that sense they 'must be there'; but they do not themselves 'make' the building in the sense of constructing it, but only in that of constituting its material. What causes the building to be made is the purpose of protecting and preserving certain goods. And so in all other cases where a purpose can be traced. It cannot be accomplished without materials that have the required nature; but it is not they that 'make' the purpose-fulfilling instrument, except materially. For what makes it, in the formative sense, is the purposeful intention of the maker. For instance: 'Why is a saw like this?' 'In order that it may have the essential character of a saw and serve for sawing.' This purpose, however, could not be served if it were not made of iron. So if it is to be a saw, and to do its work, it 'must necessarily' be made of iron. The necessity, then, is conditional, or hypothetical. The purpose, mentally conceived, demands the material as necessary to its accomplishment; but the nature of the material, as already existing, does not 'necessarily' lead to the accomplishment of the purpose.[b]

Note, too, that in a certain respect, there is a kind

[a] [γῆ, unbaked brick, of which house walls were commonly made.—C.]

[b] [*Literally*, 'So the "necessity" here is necessity in the hypothetical sense, but not in the sense in which an end (is a "necessary" consequence of its antecedents); for it resides in the matter, whereas the end is contained in the notion (or definition of the thing's essential nature, *e.g.* "what it is to be a saw").' *Cf. De part. anim.* 639 b.—C.]

of inverted parallelism between mathematics and the things that take place under the control of nature. For a straight line being what it is, if (i) a triangle is straight-sided, then (ii) its angles are equal to two right angles; but it does not follow that if (ii) its angles are equal to two right angles, then (i) the triangle must be straight-sided, though it does follow that if (ii) is not, *i.e.* its angles are not equal to two right angles, then (i) is not, *i.e.* the triangle cannot be straight-sided.[a] But in purposeful constructions it is the other way about; for if (ii) the end is to be secured or actually has been secured, then (i) the antecedent conditions must be, or must have been, present; but just as in the case of the triangle the failure of the *consequent* (its angles are equal to two right angles) necessarily involves the failure of the *antecedent* (that the triangle be straight-sided), so here the failure of the material *antecedent* (bricks or iron) necessarily carries with it the failure of the proposed *end* (house or saw). But in both cases it is the 'principle' or 'primary factor' that imposes the necessity upon the accessary one; for in the case of the triangle it is the premise that imposes a logical necessity on the conclusion (for the whole construc-

[a] [*Or*, 'There is, in a way, some resemblance between the working of (logical) necessity in mathematics and the necessity involved in things coming into existence by process of nature. Thus, if the straight line is defined as it is, it is a necessary consequence that the angles of the triangle should be equal to two right angles. On the other hand we cannot prove the truth of the definition from this property of the triangle; though we can say that, if the triangle has not this property, then nothing answering to the definition of "straight line" exists.' (The property of the (rectilinear) triangle, demonstrated in Euclid i. 32, follows from (1) the definition of straight line (and of other terms), (2) the "hypothesis" that such things as those defined *exist*, (3) other assumptions. No one of these assumptions can be proved from the consequence of them all. The ultimate principles are, in fact, indemonstrable. On the other hand, if the conclusion turned out to be false, we could infer that some hypothesis, such as the existence of the straight line, was false. *Cf. Anal. post.* i. 10; Heath, *Thirteen Books of Euclid's Elements* i. 117 ff.) 'But in the production of things for a purpose (of art or of Nature), the necessity works the other way: the assumption is that the *end* is to exist or does exist, and this necessitates that the antecedents shall exist or do exist; otherwise, as in mathematics the failure of the conclusion necessitates the non-existence of the premiss we start from (straight line), here the non-existence of the end or purpose will be necessitated. For the end or purpose also is a principle or starting-point, not indeed of the process of realization, but of the process of reasoning; whereas in mathematics there is only the starting-point, or premiss, of the reasoning, no action being involved.'

Cf. Eth. Nic. 1112 b 11 ff. In action we reason backwards from the proposed end ('If I am to bring this about, what must I do first?') along the chain of means, till we arrive at some action we can at once perform. We perform it, and so start the train of means from its ἀρχή (the beginning of the action and the conclusion of the reasoning) to the τέλος (the end of the action and the starting-point of the reasoning).—C.]

tion is logical, and there is no action at all in question), whereas in the other cases, though the purpose (which is the principle that imposes the necessity on the material) has to do with action, yet it is only a logical necessity that it can impose, for it does not actually create or constitute the material. Thus, if there is to be a house, then stones, bricks and so forth, if not already on hand, must be produced (for at any rate they must be there). Or, to put it generally, the purpose-serving materials must be there, if the purpose is to be accomplished; but since they do not themselves fulfil the purpose, except materially, it does not follow that, if they are there, the purpose will be fulfilled. To sum up: the materials will not account for the existence of the house or of the saw, though, if they are simply not there—no stones for the house, no iron for the saw—there will be no house and no saw (just as in the mathematical illustration straight-sidedness does not inhere in biorthogonality, but cannot exist without it).[a]

It is clear, then, that when physicists speak of necessity absolutely, they should limit the term to what is inherent in the material, and should recognize purposeful movement imposed upon the material as a distinct addition to its inherent qualities.[b] And though the physicist has to deal with both material and purpose, he is more deeply concerned with the latter; for purpose directs the moving causes that act upon the material, not the reverse. It is the goal that determines the purpose, and the principle of causation is derived from the definition and rationale of the end, in Nature just as much as in artificial

[a] [*Or*, '(as in the mathematical illustration, entities whose existence is assumed at the outset will not exist, if it is not true that the angles of a triangle are equal to two right angles).' Aristotle makes confusion by at once comparing and contrasting logical and teleological necessity. His real point is that logical necessity works forward from premiss to conclusion, but in Nature there is not only mechanical necessity, working blindly forward from cause to effect, but 'hypothetical' necessity, seen in the processes of Nature or of art, where a 'final cause' fixes the goal to which the mechanical chain shall lead, rather than in other directions. The assumption, by the artist or by Nature, that this end shall be brought about, starts the whole process and accounts for its course.—C.]

[b] [*Literally* 'what is necessary in the things of Nature is what is so called in the sense of (the necessary) matter and its motions' (its inherent behaviour). These are a condition *sine qua non*, but will not of themselves produce the desirable 'end.'—C.]

constructions. *E.g.* since this is what a house is, these things must necessarily be ready or be produced; and since this is what health is, such and such things must be provided to secure it. And in like manner, if this is what 'man' is, then there must antecedently be such and such things, and that they again may be there, such and such other things. And the same kind of 'necessity' may be traced on the conceptual side.[a] For if we define 'sawing' as such and such a method of dividing, it follows that it can only be accomplished by teeth of such and such a character, and these teeth can exist only if the saw be made of iron. For a definition, too, contains constituent terms which are, as it were, its materials.

[a] [*Or*, 'Though, perhaps, we may also say that what is "necessary" in this way is actually contained in the definition.']

2

Reprinted from pages 246–252 of *De Rerum Natura, Libri Sex,* vol. 1, H. A. J. Munro, trans., Deighton, Bell and Co., Cambridge, England, 1864

DE RERUM NATURA, LIBRI SEX

Lucretius

[*Editor's Note:* In the original, material precedes and follows this excerpt. The Latin text, which appears in the original, has been omitted.]

In the beginning the earth gave forth all kinds of herbage and verdant sheen about the hills and over all the plains; the flowery meadows glittered with the bright green hue, and to the different trees was given a strong and emulous desire of growing up into the air with full unbridled powers. As feathers and hairs and bristles are first born on the limbs of four-footed beasts and the body of the strong of wing, thus the new earth then first put forth grass and bushes, and next gave birth to the races of mortal creatures springing up many in number in many ways after diverse fashions. For no living creatures can have dropped from heaven nor can those belonging to the land have come out of the salt pools. It follows that with good reason the earth has gotten the name of mother, since all things have been produced out of the earth. And many living creatures even now spring out of the earth taking form by rains and the heat of the sun. It is therefore the less strange if at that time they sprang up more in number and larger in size, having come to maturity in the infancy of earth and ether. First of all the race of fowls and the various birds would leave their eggs, hatched in the springtime, just as now in summer the cicades leave spontaneously their delicate coats in quest of a living and life. Then you must know did the earth first give forth races of mortal men. For much heat and moisture would then abound in the fields; and therefore wherever a suitable spot offered, wombs would grow attached to the earth by roots; and when the warmth of the infants, flying the wet and craving the air, had opened these in the fulness of time, nature would turn to that spot the pores of the earth and constrain it to yield from its opened veins a liquid most like to milk, even as now-a-days every woman when she has borne, is filled with sweet milk, because all that current of nutriment streams towards the breasts. To the children the earth would furnish food, the heat raiment, the grass a bed rich in abundance of soft down. But the fresh youth of the world would give forth neither severe colds nor excessive heats nor gales of great violence; for all things grow and acquire strength in a like proportion.

Wherefore again and again I say the earth with good title has gotten and keeps the name of mother, since she of herself gave birth to mankind and at a time nearly fixed shed forth every beast that ranges wildly over the great mountains, and at the same time the fowls of the air with all their varied shapes. But because she must have some limit set to her bearing, she ceased like a woman worn out by length of days. For time changes the nature of the whole world and all things must pass on from one condition to another, and nought continues like to itself: all things quit their bounds, all things nature changes and compels to alter. One thing crumbles away and is worn and enfeebled with age, then another comes unto honour and issues out of its state of contempt. In this way then time changes the nature of the whole world and the earth passes out of one condition into another, so that now it cannot bear what once it could, and now it can bear what before it did not bear.

And many monsters too the earth at that time essayed to produce, things coming up with with strange face and limbs, the man-woman, a thing between the two and neither the one sex nor the other, widely differing from both; some things deprived of feet, others again destitute of hands, others too proving dumb without mouth, or blind without eyes, and things bound fast by the adhesion of their limbs over all the body, so that they could not do anything nor go anywhere nor avoid the evil nor take what their needs required. Every other monster and portent of this kind she would produce, but all in vain, since nature set a ban on their increase and they could not reach the coveted flower of age nor find food nor be united in marriage. For we see that many conditions must meet together in things in order that they may beget and continue their kinds; first a supply of food, then a way by which the birth-producing seeds throughout the frame may stream from the relaxed limbs; also in order that the woman may be united with the male, the possession of organs whereby they may each interchange mutual joys.

And many races of living things must then have died out and been unable to beget and continue their breed. For in the case of all things which you see breathing the breath of life, either craft or courage or else speed has from the beginning of its existence protected and preserved each particular race. And there are many things which, recommended to us by their useful services, continue to exist consigned to our protection. In the first place the fierce breed of lions and the savage races their courage has protected, foxes their craft and stags their proneness to flight. But light-sleeping dogs with faithful heart in breast and every kind which is born of the seed of beasts of burden and at the same time the woolly flocks and the horned herds are all consigned, Memmius, to the protection of man. For they have ever fled with eagerness from wild-beasts and have ensued peace, and plenty of food has been obtained without their own labour, as we give it in requital of their useful services. But those to whom nature has granted none of these qualities, so that they could neither live by their own means nor perform for us any useful service in return for which we should suffer their kind to feed and be safe under our protection, those, you are to know, would lie exposed as a prey and booty of others, hampered all in their own death-bringing shackles, until nature brought that kind to utter destruction.

But Centaurs never have existed, and at no time can there exist things of twofold nature and double body formed into one frame out of limbs of alien kinds, such that the faculties and powers of this and that portion cannot be sufficiently like. This however dull of understanding you may learn from what follows. To begin, a horse when three years have gone round is in the prime of his vigour, far different the boy: often even at that age he will call in his sleep for the milk of the breast. Afterwards when in advanced age his lusty strength and limbs now faint from ebbing life fail the horse, then and not till then youth in the flower of age commences for that boy and clothes his cheeks in soft down; that you may not haply believe that out of a man and the burden-carrying seed of horses Centaurs can be formed and have being; or that Scyllas with bodies half those of fishes girdled round with raving dogs can exist, and all other things of the kind, whose limbs we see cannot harmonise together; as they neither come to their flower at the same time nor reach the fulness of their bodily strength nor lose it in advanced old age, nor burn with similar passions nor have compatible manners, nor feel the same things give pleasure throughout their frames. Thus we may see bearded goats often fatten on hemlock which for man is rank poison. Since flame moreover is wont to scorch and burn the tawny bodies of lions just as much as any other kind of flesh and blood existing on earth, how could it be that a single chimera with triple body, in front a lion, behind a dragon, in the middle the goat whose name it bears, could breathe out at the mouth fierce flame from its body? Wherefore also he who fables that in the new time of the earth and the fresh youth of heaven such living creatures could have been begotten, resting upon this one futile term new, may babble out many things in like fashion, may say that rivers ran with gold over all parts of the earth, and that trees were wont to blossom with precious stones, or that man was born with such giant force of frame that he could wade on foot through deep seas and whirl the whole heaven about him with his hands. For the fact that there were many seeds of things in the earth what time it first shed forth living creatures, is yet no proof that there could have been produced beasts of different kinds mixed together, and limbs of different living things formed into a single frame, because the kinds of herbage and corn and joyous trees which now also spring in plenty out of the earth yet cannot be produced with the several sorts plaited together into one whole, but each thing goes on after its own fashion, and all preserve their distinctive differences according to a fixed law of nature.

Part II

NATURAL THEOLOGY—THE ARGUMENT FROM DESIGN

Editor's Comments on Papers 3, 4, and 5

3A **LACTANTIUS**
Of the Production of the Beasts and of Man

3B **LACTANTIUS**
Of the Error of Epicurus, and of the Limbs and Their Use

4A **PALEY**
Excerpt from *Application of the Argument Continued*

4B **PALEY**
Excerpt from *The Goodness of the Deity*

5 **WHEWELL**
Excerpt from *The Length of the Year*

The speculative scientific inquiry that began in ancient Greece went into eclipse and did not flourish again in the western world until the Renaissance—over 1500 years later. Christians cited divine authority as revealed in the Bible and interpreted by the Church fathers as the source of all knowledge. This attitude, coupled with their persecution of dissident thinkers, did not encourage scientific inquiry as a means of gaining new knowledge about the world. Speculative thinking about the origin and evolution of species was particularly inhibited by the Christian belief in the divine authority of the Old Testament book of Genesis and in its interpreters, the Church fathers. Nonetheless, the scientific revolution that led to modern science began in a Europe dominated by the Judeo-Christian religious tradition.

Scholars used teleology—the Aristotelian doctrine of final causes—to argue from the existence of adaptations in nature to a God who produced organisms with their adaptations by special acts of creation. The doctrine of final causes provided scholars with an emotionally satisfying explanation of adaptations. The historian Conway Zirkle (1941, p. 113) contended that the widespread acceptance of teleology made a scientific theory such as "natural selection an unneeded hypothesis and probably served as a protective inoculation against the spread of the doctrine."

The first Christian scholar to develop an extensive argument from the existence of design in nature to the existence of a God who created the design wrote at the beginning of the fourth century A.D. In *On the Workmanship of God,* Carius Lactantius (240–320) interpreted nature in terms of a divine providence (Paper 3A), attacked the views of Epicurus and Lucretius on the origin of species (Paper 3B), and developed the argument for a divine designer from the existence of design in nature using the analogy of a building and its designer, an architect (Paper 3B).

Many scientists—Newton, Ray, Linnaeus, Bonnet, and Cuvier for example—embraced both the scientific paradigm for learning about nature and the Biblical paradigm with its revelations about nature (Greene, 1959). Tremendous gains in biological and geological knowledge were achieved by these and other supporters of a Biblical paradigm who contended that they could also consistently use the rules of scientific analysis to gain knowledge about nature (Hooykaas, 1972; Jaki, 1978; Raven, 1953). When scientific investigations produced anomalies that did not fit a literal Biblical paradigm, few scientists completely abandoned the special creation paradigm. Instead they attempted to defend the essence of the Biblical special creation paradigm by adding ad hoc hypotheses to it. For example, the Noachian flood was no longer viewed as the only global catastrophe to have occurred—there were many global catastrophies each involving mass extinctions and each followed by new acts of divine creation.

This simultaneous commitment by many scholars to both the Biblical and the scientific paradigms with their different rules for the acquisition of knowledge—the appeal to divine revelation contained in the Bible versus the scientific appeal to objective experience as a means for testing theories—played an important role in setting the stage for the triumph of the evolutionary paradigm over the special creation paradigm among intellectuals (Ruse, 1970).

The English theologian and utilitarian philosopher William Paley (1743–1805) argued in his *Natural Theology* (Paley, 1802) that if the existence of a watch (design) is proof of a watchmaker, then the existence of contrivances (adaptive design) in nature is evidence for the existence of a divine contriver. While a student at Cambridge University, Charles Darwin read Paley's *Natural Theology* and on the eve of the publication of the *Origin of Species* wrote: "I do not think I hardly ever admired a book more than Paley's *Natural Theology.* I could almost formerly have said it by heart" (Darwin, 1887, II, p. 5).

Paley summarized and attacked the naturalistic theories concerning the origin of adaptations championed by Lucretius (Paper 4A). Buffon's theories concerning the role of generation, "organic molecules," and "internal moulds" were discussed and rejected by Paley (1802, pp. 459–463). He also summarized and rejected Erasmus Darwin's theory that adaptation was the outcome of the inheritance of characteristics acquired by passive and active habits as well as by "appentencies" (Paley, 1802, pp. 463–473). Reverend Paley also discussed Malthusian population theory, especially the destruction that follows from the overproduction of offspring (Paper 4B). Paley attempted "to point out the advantages which are gained by the powers in nature of a superabundant multiplication" (Paper 4B, p. 510) in order to counteract the argument that the animal hostilities produced by such superfecundity forms an instance "in the works of the Deity, of an oeconomy, stamped by marks of design, in which the character of utitlity can be called into question" (Paper 4B, p. 516).

In 1829 the Earl of Bridgewater died leaving a will that allocated funds for the publication of a work "On the Power, Wisdom and Goodness of God as manifested in the Creation; illustrating such work by all reasonable arguments, as for instance the variety and formation of God's Creatures in the animal, vegetable and mineral kingdoms; the effect of digestion and thereby of conversion; the construction of the hand of man and an infinite variety of other arguments; as also by discoveries ancient and modern, in arts, science and the whole extent of literature" (Raven, 1953, p. 210). The work, which was published in eight volumes during the 1830s, became known as the Bridgewater Treatises. The volume Astronomy and Physics, written by the English astronomer and philosopher of science William Whewell (1794–1866), contains a passage summarizing and rejecting a theory of gradual evolution of adaptations produced by selection operating on chance variations (Paper 5). Whewell's use of the argument from design in his Bridgewater Treatise and subsequent writings has been analyzed by Ruse (1977).

The theological argument from design in nature was based on teleological thinking that explained organic structure and function by reference to purpose or final causes. The role that such arguments played in the development of Charles Darwin's theory of "descent with modification" by natural, sexual, and artificial selection has been recently reviewed by Ruse (1979) and Ospovat (1981).

REFERENCES

Darwin, F., ed., 1887, 1899, *The Life and Letters of Charles Darwin,* 2 vols., Appleton, New York, 558p., 562p.

Greene, J. C., 1959, *The Death of Adam: Evolution and Its Impact on Western Thought,* Iowa State University Press, Ames, Iowa, 388p.

Hooykaas, R., 1972, *Religion and the Rise of Modern Science,* Eerdmans, Grand Rapids, Michigan, 162p.

Jaki, S., 1978, *The Road of Science and the Ways to God,* University of Chicago Press, Chicago, 478p.

Ospovat, D., 1981, *The Development of Darwin's Theory: Natural History, Natural Theology, and Natural Selection, 1838-1859,* Cambridge University Press, Cambridge, 301p.

Paley, W., 1802, *Natural Theology: or, Evidences of the Existence and Attributes of The Deity, Collected from the Appearances of Nature,* R. Faulder, London, 586p. (Reprinted in facsimile in 1970 by Gregg International, Westmead, England.)

Raven, C., 1953, *Natural Religion and Christian Theology,* Cambridge University Press, Cambridge, 224p.

Ruse, M., 1979, *Darwinian Revolution: Nature Red in Tooth and Claw,* University of Chicago Press, Chicago, 320p.

Ruse, M., 1977, William Whewell and the Argument from Design *Monist* **60:**244-268.

Zirkle, C., 1941, Natural Selection Before the "Origin of Species," *Am. Philos. Soc. Proc.* **84:**71-123.

3A

Reprinted from pages 51-53 of *The Works of Lactantius,* vol. 2, W. Fletcher, trans., T. & T. Clark, Edinburgh, 1886

OF THE PRODUCTION OF THE BEASTS AND OF MAN

Lactantius

For our Creator and Parent, God, has given to man perception and reason, that it might be evident from this that we are descended from Him, because He Himself is intelligence, He Himself is perception and reason. Since He did not give that power of reason to the other animals, He provided beforehand in what manner their life might be more safe. For He clothed them all with their own natural hair,[1] in order that they might more easily be able to endure the severity of frosts and colds. Moreover, He has appointed to every kind its own peculiar defence for the repelling of attacks from without; so that they may either oppose the stronger animals with natural weapons, or the feebler ones may withdraw themselves from danger by the swiftness of their flight, or those which require at once both strength and swiftness may protect themselves by craft, or guard themselves in hiding-places. And so others of them either poise themselves aloft with light plumage, or are supported by hoofs, or are furnished with horns; some have arms in their mouth—namely, their teeth—or hooked talons on their feet; and none of them is destitute of a defence for its own protection.

But if any fall as a prey to the greater animals, that their race might not utterly perish, they have either been banished to that region where the greater ones cannot exist, or they have received a more abundant fruitfulness in production, that food might be supplied from them to the beasts which are nourished by blood, and yet their very multitude might survive the slaughter inflicted upon them, so as to preserve the race. But He made man—reason being granted to him, and the power of perceiving and speaking being given to him—destitute of those things which are given to the other animals, because wisdom was able to supply those things which the condition of nature had denied to him. He made him naked and defenceless, because he could be armed by his talent, and clothed by his reason. But it cannot be expressed how wonderfully the absence of those things which are given to the brutes contributes to the beauty of man. For if He had given to man the teeth of wild beasts, or horns, or claws, or hoofs, or a tail, or hairs of various colour, who cannot perceive how misshapen an animal he would be, as the dumb animals, if they were made naked and defenceless? For if you take from these the natural clothing of their body, or those things by which they are armed of themselves, they can be neither beautiful nor safe, so that they appear wonderfully furnished if you think of utility, and wonderfully adorned if you think of appearance: in such a wonderful manner is utility combined with beauty.

But with reference to man, whom He formed an eternal and

[1] "Omnes enim suis ex se pilis." Others read, "pellibus texit."

immortal being, He did not arm him, as the others, without, but within; nor did He place his protection in the body, but in the soul: since it would have been superfluous, when He had given him that which was of the greatest value, to cover him with bodily defences, especially when they hindered the beauty of the human body. On which account I am accustomed to wonder at the senselessness of the philosophers who follow Epicurus, who blame the works of nature, that they may show that the world is prepared and governed by no providence; but they ascribe the origin of all things to indivisible and solid bodies, from the fortuitous meetings of which they say that all things are and were produced. I pass by the things relating to the world itself with which they find fault, in which matter they are ridiculously mad; I assume that which belongs to the subject of which we are now treating.

3B

Reprinted from pages 62-64 of *The Works of Lactantius,* vol. 2, W. Fletcher, trans., T. & T. Clark, Edinburgh, 1886

OF THE ERROR OF EPICURUS, AND OF THE LIMBS AND THEIR USE

Lactantius

I cannot here be prevented from again showing the folly of Epicurus. For all the ravings of Lucretius belong to him, who, in order that he might show that animals are not produced by any contrivance of the divine mind, but, as he is wont to say, by chance, said that in the beginning of the world innumerable other animals of wonderful form and magnitude were produced; but that they were unable to be permanent, because either the power of taking food, or the method of uniting and generating, had failed them. It is evident that, in order to make a place for his atoms flying about through the boundless and empty space, he wished to exclude the divine providence. But when he saw that a wonderful system of providence is contained in all things which breathe, what vanity was it (O mischievous one!) to say that there had been animals of immense size, in which the system of production ceased!

Since, therefore, all things which we see are produced with reference to a plan—for nothing but a plan[1] can effect this very condition of being born—it is manifest that nothing could have been born without a plan. For it was previously foreseen in the formation of everything, how it should use the service of the limbs for the necessaries of life; and how the offspring, being produced from the union of bodies, might preserve all living creatures by their several species. For if a skilful architect, when he designs to construct some great building, first of all considers what will be the effect[2] of the complete building, and previously ascertains by measurement what situation is suitable for a light weight, in what place a massive part of the structure will stand, what will be the intervals between the columns, what or where will be the descents and outlets of the falling waters and the reservoirs,—he first, I say, foresees these things, that he

[1] "Ratio." Nearly equivalent in this place to "providentia."

[2] "Summa."

may begin together with the very foundations whatever things are necessary for the work when now completed,—why should any one suppose that, in the contrivance of animals, God did not foresee what things were necessary for living, before giving life itself? For it is manifest that life could not exist, unless those things by which it exists were previously arranged.

Therefore Epicurus saw in the bodies of animals the skill of a divine plan; but that he might carry into effect that which he had before imprudently assumed, he added another absurdity agreeing with the former. For he said that the eyes were not produced for seeing, nor the ears for hearing, nor the feet for walking, since these members were produced before there was the exercise of seeing, hearing, and walking; but that all the offices of these members arose from them after their production. I fear lest the refutation of such extravagant and ridiculous stories should appear to be no less foolish; but it pleases me to be foolish, since we are dealing with a foolish man, lest he should think himself too clever. What do you say, Epicurus? Were not the eyes produced for seeing? Why, then, do they see? Their use, he says, afterwards showed itself. Therefore they were produced for the sake of seeing, since they can do nothing else but see. Likewise, in the case of the other limbs, use itself shows for what purpose they were produced. For it is plain that this use could have no existence, unless all the limbs had been made with such arrangement and foresight, that they might be able to have their use.

For what if you should say, that birds were not made to fly, nor wild beasts to rage, nor fishes to swim, nor men to be wise, when it is evident that living creatures are subject to that natural disposition and office to which each was created? But it is evident that he who has lost the main point itself of the truth must always be in error. For if all things are produced not by providence, but by a fortuitous meeting together of atoms, why does it never happen by chance, that those first principles meet together in such a way as to make an animal of such a kind, that it might rather hear with its nostrils, smell with its eyes, and see[1] with its ears? For if the first principles leave no kind of position untried, monstrous productions of this kind ought daily to have been brought forth, in

[1] "Cerneret," to see so as to distinguish; a stronger word than "video."

which the arrangement of the limbs might be distorted,[1] and the use far different from that which prevails. But since all the races of animals, and all the limbs, observe their own laws and arrangements, and the uses assigned to them, it is plain that nothing is made by chance, since a perpetual arrangement of the divine plan is preserved. But we will refute Epicurus at another time. Now let us discuss the subject of providence, as we have begun.

[1] "Præposterus;" having the last first, and the first last.

4A

Reprinted from pages 68–72 of *Natural Theology; or, Evidences of the Existence and Attributes of the Deity, Collected from the Appearances of Nature*, Faulder, London, 1802

APPLICATION OF THE ARGUMENT CONTINUED

W. Paley

[*Editor's Note:* In the original, material precedes and follows this excerpt.]

IV. There is another anſwer which has the ſame effect as the reſolving of things into chance; which anſwer would perſuade us to believe, that the eye, the animal to which it belongs, every other animal, every plant, indeed every organized body which we ſee, are only ſo many out of the poſſible varieties and combinations of being, which the lapſe of infinite ages has brought into exiſtence; that the preſent world is the relict of that variety; millions of other bodily forms and other ſpecies having periſhed, being by the defect of their conſtitution incapable of preſervation, or of continuance by generation. Now there is no foundation whatever for this conjecture in any thing which we obſerve in the works of nature;

ture: no ſuch experiments are going on at preſent; no ſuch energy operates as that which is here ſuppoſed, and which ſhould be conſtantly puſhing into exiſtence new varieties of beings: nor are there any appearances to ſupport an opinion, that every poſſible combination of vegetable or animal ſtructure has formerly been tried. Multitudes of conformations, both of vegetables and animals, may be conceived capable of exiſtence and ſucceſſion, which yet do not exiſt. Perhaps almoſt as many forms of plants might have been found in the fields, as figures of plants can be delineated upon paper. A countleſs variety of animals might have exiſted which do not exiſt. Upon the ſuppoſition here ſtated, we ſhould ſee unicorns and mermaids, ſylphs and centaurs; the fancies of painters and the fables of poets realized by examples. Or, if it be alledged that theſe may tranſgreſs the limits of poſſible life and propagation, we might, at leaſt, have nations of human beings without nails upon their fingers, with more or fewer fingers and toes than ten, ſome with one eye, others with one ear, with one noſtril, or without the ſenſe of ſmelling at all. All theſe, and a thouſand other imaginable varieties, cties, might live and propagate. We may modify any one ſpecies many different ways, all conſiſtent with life, and with the actions neceſſary to preſervation, although affording different degrees of conveniency and enjoyment to the animal. And if we carry theſe modifications through the different ſpecies which are known to ſubſiſt, their number would be incalculable. No reaſon can be given why, if theſe deperdits ever exiſted, they have now diſappeared. Yet, if all poſſible exiſtences have been tried, they muſt have formed part of the catalogue.

But, moreover, the diviſion of organized ſubſtances into animals and vegetables, and the diſtribution and ſub-diſtribution of each into genera and ſpecies, which diſtribution is not an arbitrary act of the mind, but is founded in the order which prevails in external nature, appear to me to contradict the ſuppoſition of the preſent world being the remains of an indefinite variety of exiſtences; of a variety which rejects all plan. The hypotheſis teaches, that every poſſible variety of being hath, at one time or other, found its way into exiſtence by what cauſe or in what manner is not ſaid), and that thoſe which were badly formed,

formed, perifhed: but how or why thofe which furvived fhould be caft, as we fee that plants and animals are caft, into regular claffes, the hypothefis does not explain; or rather the hypothefis is inconfiftent with this phænomenon.

The hypothefis, indeed, is hardly deferving of the confideration which we have given to it. What fhould we think of a man, who, becaufe we had never ourfelves feen watches, telefcopes, ftocking-mills, fteam-engines, &c. made; knew not how they were made; or could prove by teftimony when they were made, or by whom;—would have us believe that thefe machines, inftead of deriving their curious ftructures from the thought and defign of their inventors and contrivers, in truth derive them from no other origin than this; that, a mafs of metals and other materials having run when melted into all poffible figures, and combined themfelves in all poffible forms and fhapes and proportions, thefe things which we fee, are what were left from the accident, as beft worth preferving; and, as fuch, are become the remaining ftock of a magazine, which, at one time or other, has, by this means, contained every mechanifm, ufeful and ufelefs, convenient and inconvenient, into which fuch like materials could be thrown? I cannot diftinguifh the hypothefis as applied to the works of nature, from this folution, which no one would accept, as applied to a collection of machines.

4B

Reprinted from pages 510–516 of *Natural Theology; or, Evidences of the Existence and Attributes of the Deity, Collected from the Appearances of Nature,* Faulder, London, 1802

THE GOODNESS OF THE DEITY

W. Paley

[*Editor's Note:* In the original, material precedes and follows this excerpt.]

3. But, to do juftice to the queftion, the fyftem of animal *deftruction* ought always to be confidered in ftrict connection with another property of animal nature, viz. *fuperfecundity* They are countervailing qualities. One fubfifts by the correction of the other. In treating, therefore, of the fubject under this view, (which is, I believe, the true one,) our bufinefs will be, firft, to point out the advantages which are gained by the powers in nature of a fuperabundant multiplication; and, then, to fhew, that thefe advantages are fo many reafons for appointing that fyftem of animal hoftilities, which we are endeavouring to account for.

In almoft all cafes nature produces her fupplies with profufion. A fingle cod fifh fpawns, in one feafon, a greater number of eggs, than all the inhabitants of England amount to. A thoufand other inftances of prolific generation might be ftated, which, though not equal to this, would carry on the increafe of the fpecies with a rapidity which outruns calculation, and to an immeafurable extent. The advantages of fuch a conftitution are two: firft, that it tends to keep the world always full; whilft, fecondly, it allows the proportion between the

feveral

ſeveral ſpecies of animals to be differently modified, as different purpoſes require, or as different ſituations may afford for them room and food. Where this vaſt fecundity meets with a vacancy fitted to receive the ſpecies, there it operates with its whole effect; there it pours in its numbers, and repleniſhes the waſte. We complain of what we call the exorbitant multiplication of ſome troubleſome inſects, not reflecting that large portions of nature might be left void without it. If the accounts of travellers may be depended upon, immenſe tracts of foreſt in North America would be nearly loſt to ſenſitive exiſtence if it were not for *gnats*. "In the thinly inhabited regions of America, in which the waters ſtagnate, and the climate is warm, the whole air is filled with crowds of theſe inſects." Thus it is, that, where we looked for ſolitude and death-like ſilence, we meet with animation, activity, enjoyment; with a buſy, a happy, and a peopled world. Again; hoſts of *mice* are reckoned amongſt the plagues of the north-eaſt part of Europe; whereas vaſt plains in Siberia, as we learn from good authority, would be lifeleſs without them. The Caſpian deſarts are converted by their preſence into crowded warrens.

warrens. Between the Volga and the Yaik, and in the country of Hyrcania, the ground, ſays Pallas, is in many places *covered* with little hills, raiſed by the earth caſt out in forming the burrows. Do we then ſo envy theſe bliſsful abodes, as to pronounce the fecundity by which they are ſupplied with inhabitants, to be an evil; a ſubject of complaint, and not of praiſe? Further; by virtue of this ſame ſuperfecundity, what we term deſtruction, becomes almoſt inſtantly the parent of life. What we call blights, are, oftentimes, legions of animated beings claiming their portion in the bounty of nature. What corrupts the produce of the earth to us, prepares it for them. And it is by means of their rapid multiplication, that they take poſſeſſion of their paſture: a ſlow propagation would not meet the opportunity.

But in conjunction with the occaſional uſe of this fruitfulneſs, we obſerve, alſo, that it allows the proportion between the ſeveral ſpecies of animals to be differently modified, as different purpoſes of utility may require. When the foreſts of America come to be cleared, and the ſwamps drained, our gnats will give place to other inhabitants. If the population of Europe ſhould ſpread to the north

north and the eaſt, the mice will retire before the huſbandman and the ſhepherd, and yield their ſtation to herds and flocks. In what concerns the human ſpecies, it may be a part of the ſcheme of Providence that the earth ſhould be inhabited by a ſhifting, or perhaps a circulating population. In this œconomy it is poſſible that there may be the following advantages. When old countries are become exceedingly corrupt, ſimpler modes of life, purer morals, and better inſtitutions may riſe up in new ones, whilſt freſh ſoils reward the cultivator with more plentiful returns. Thus the different portions of the globe come into uſe in ſucceſſion as the reſidence of man; and, in his abſence, entertain other gueſts, which, by their rapid multiplication ſoon fill the chaſm. In domeſticated animals we find the effect of their fecundity to be, that we can always command *numbers*: we can always have as many of any particular ſpecies as we pleaſe, or as we can ſupport. Nor do we complain of its exceſs; it being much more eaſy to regulate abundance, than to ſupply ſcarcity.

But then this *ſuperfecundity*, though of great occaſional uſe and importance, exceeds the ordinary capacity of nature to receive or ſupport

ſupport its progeny. All ſuperabundance ſuppoſes deſtruction, or muſt deſtroy itſelf. Perhaps there is no ſpecies of terreſtrial animals whatever, which would not overrun the earth, if it were permitted to multiply in perfect ſafety; or of fiſh, which would not fill the ocean: at leaſt, if any ſingle ſpecies were left to their natural increaſe without diſturbance or reſtraint, the food of other ſpecies would be exhauſted by their maintenance. It is neceſſary, therefore, that the effects of ſuch prolific faculties be curtailed. In conjunction with other checks and limits, all ſubſervient to the ſame purpoſe, are the thinnings which take place among animals, by their action upon one another. In ſome inſtances we ourſelves experience, very directly, the uſe of theſe hoſtilities. One ſpecies of inſects rids us of another ſpecies; or reduces their ranks. A third ſpecies perhaps keeps the ſecond within bounds: and birds or lizards are a fence againſt the inordinate increaſe by which even theſe laſt might infeſt us. In other, more numerous, and poſſibly more important inſtances, this diſpoſition of things, although leſs neceſſary or uſeful to us, and of courſe leſs obſerved by us, may be neceſſary and uſeful to certain other

ſpecies;

fpecies; or even for the preventing of the lofs of certain fpecies from the univerfe: a misfortune which feems to be ftudioufly guarded againft. Though there may be the appearance of failure in fome of the details of Nature's works, in her great purpofes there never are. Her fpecies never fail. The provifion which was originally made for continuing the replenifhment of the world has proved itfelf to be effectual through a long fucceffion of ages.

What further fhews, that the fyftem of deftruction amongft animals holds an exprefs relation to the fyftem of fecundity; that they are parts indeed of one compenfatory fcheme; is, that, in each fpecies, the fecundity bears a proportion to the fmallnefs of the animal, to the weaknefs, to the fhortnefs of its natural term of life, and to the dangers and enemies by which it is furrounded. An elephant produces but one calf: a butterfly lays fix hundred eggs. Birds of prey feldom produce more than two eggs: the fparrow tribe, and the duck tribe, frequently fit upon a dozen. In the rivers, we meet with a thoufand minnows for one pike; in the fea, a million of herrings for a fingle fhark. Compenfation obtains

obtains throughout. Defencelefsnefs and devaftation are repaired by fecundity.

We have dwelt the longer upon thefe confiderations, becaufe the fubject to which they apply, namely, that of animals *devouring* one another, forms the chief, if not the only inftance, in the works of the Deity, of an œconomy, ftamped by marks of defign, in which the character of utility can be called in queftion. The cafe of *venomous* animals is of much inferior confequence to the cafe of prey, and, in fome degree, is alfo included under it. To both cafes it is probable that many more reafons belong, than thofe of which we are in poffeffion.

Our *firft* propofition, and that which we have hitherto been defending, was, "that in a vaft plurality of inftances, in which *contrivance* is perceived, the defign of the contrivance is beneficial."

Our *fecond* propofition is, "that the Deity has added *pleafure* to animal fenfations, beyond what was neceffary for any other purpofe, or when the purpofe, fo far as it was neceffary, might have been effected by the operation of pain."

5

Reprinted from pages 26–28 of *Astronomy and General Physics, Considered with General Reference to Natural Theology,* Treatise III in The Bridgewater Treatises on the Power, Wisdom, and Goodness of God, as Manifested in Creation, vol. 2, new ed., Carey, Lea & Blanchard, Philadelphia, 1836

THE LENGTH OF THE YEAR

W. Whewell

[*Editor's Note:* In the original, material precedes this excerpt.]

If therefore the duration of the seasons were much to change, the processes of vegetable life would be interrupted, deranged, distempered. What, for instance, would become of our calendar of Flora, if the year were lengthened or shortened by six months? Some of the dates would never arrive in the one case, and the vegetable processes which mark them would be superseded; some seasons would be without dates in the other case, and these periods would be employed in a way harmful to the plants, and no doubt speedily destructive. We should have not only *a year of confusion*, but, if it were repeated and continued, a year of death.

But in the existing state of things, the duration of the earth's revolution round the sun, and the duration of the revolution of the vegetable functions of most plants are equal. These two periods are *adjusted* to each other. The stimulants which the elements apply come at such intervals and continue for such times, that the plant is supported in health and vigour, and enabled to reproduce its kind. Just such a portion of time is measured out for the vegetable powers to execute their task, as enables them to do so in the best manner.

Now such an adjustment must surely be accepted as a proof of design, exercised in the formation of the world. Why should the solar year be so long and no longer? or, this being of such a length, why should the vegetable cycle be exactly of the same length? Can this be chance? And this occurs, it is to be observed, not in one, or in a few species of plants, but in thousands. Take a small portion only of known species, as the most obviously endowed with this adjustment, and say ten thousand. How should all these organized bodies be constructed for the same period of a year. How should all these machines be wound up so as to go for the same time? Even allowing that they could bear a year of a month longer or shorter, how do they all come within such limits? No chance could produce such a result. And if not by chance, how otherwise could such a coincidence occur, than by an intentional adjustment of these two things to one another? by a selection of such an organization in plants, as would fit them to the earth on which they were to grow; by an adaptation of construction to conditions; of the scale of construction to the scale of conditions.

It cannot be accepted as an explanation of this fact in the economy of plants, that it is necessary to their existence; that no plants could possibly have subsisted, and come down to us, except those which were thus suited to their place on the earth. This is true; but this does not at all remove the necessity of recurring to design as the origin of the construction by which the existence and continuance of plants is made possible. A watch could not go, except there were the most exact adjustment in the forms and positions of its wheels; yet no one would accept it as an explanation of the origin of such forms and positions, that the watch would not go if these were other than they are. If the objector were to suppose that plants were originally fitted to years of various lengths, and that such only have survived to the present time, as had a cycle of a length equal to our present year, or one which could be accommodated to it; we should reply, that the assumption is too gratuitous and extravagant to require much consideration; but that, moreover, it does not remove the difficulty. How came the functions of plants to be *periodical* at all? Here is, in the first instance, an agreement in the form of the laws that prevail in the organic and in the inorganic world, which appears to us a clear evidence of design in their Author. And the same kind of reply might be made to any similar objection to our argument. Any supposition that the universe has gradually approximated to that state of harmony among the operations of its different parts, of which we have one instance in the coincidence now under consideration, would make it necessary for the objector to assume a previous state of things preparatory to this perfect correspondence. And in this preparatory condition we should still be able to trace the rudiments of that harmony, for which it was proposed to account: so that even the most unbounded licence of hypothesis would not enable the opponent to obliterate the traces of an intentional adaptation of one part of nature to another.

Nor would it at all affect the argument, if these periodical occurrences could be traced to some proximate cause: if for instance it could be shown, that the budding or flowering of plants is brought about at particular intervals, by the nutriment accumulated in their vessels during the preceding months. For the question would still remain, how their functions were so adjusted, that the accumulation of the nutriment necessary for budding and flowering, together with the operation itself, comes to occupy exactly a year, instead of a month only, or ten years. There must be in their structure some reference to time: how did such a reference occur? how was it determined to the particular time of the earth's revolution round the sun? This could be no otherwise, as we conceive, than by design and appointment.

We are left therefore with this manifest adjustment before us, of two parts of the universe, at first sight so remote; the dimensions of the solar system and the powers of vegetable life. These two things are so related, that one has been made to fit the other. The relation is as clear as that of a watch to a sundial. If a person were to compare the watch with the dial, hour after hour, and day after day, it would be impossible for him not to believe that the watch had been *contrived* to accommodate itself to the solar day. We have at least ten thousand kinds of vegetable watches of the most various forms, which are all accommodated to the solar year; and the evidence of contrivance seems to be no more capable of being eluded in this case than in the other.

The same kind of argument might be applied to the animal creation. The pairing, nesting, hatching, fledging, and flight of birds, for instance, occupy each its peculiar time of the year; and, together with a proper period of rest, fill up the twelve months. The transformations of most insects have a similar reference to the seasons, their progress and duration. "In every species" (except man), says a writer* on animals, "there is a particular period of the year in which the reproductive system exercises its energies. And the season of love and the period of gestation are so arranged that the young ones are produced at the time wherein the conditions of temperature are most suited to the commencement of life." It is not our business here to consider the details of such provisions, beautiful and striking as they are. But the prevalence of the great law of periodicity in the vital functions of organized beings will be allowed to have a claim to be considered in its reference to astronomy, when it is seen that their periodical constitution derives its use from the periodical nature of the motions of the planets round the sun; and that the duration of such cycles in the existence of plants and animals has a reference to the arbitrary elements of the solar system: a reference which, we maintain, is inexplicable and unintelligible, except by admitting into our conceptions an Intelligent Author, alike of the organic and inorganic universe.

* Fleming, Zool. i. 400.

Part III

THE FRENCH TRANSFORMISTS

Editor's Comments on Papers 6 and 7

The French mathematician and astronomer Pierre-Louis Moreau de Maupertuis (1698–1759) was the first modern scientist to consider seriously the theory that selection might play a major role in the origin of species (Glass, 1959, Bajema, 1982). In his *Venus Physique* (1745), Maupertuis contended that chance produces new varieties and he drew attention to the role that selective animal breeders play in bringing about the appearance of new races of dogs, pigeons, and canaries that had never existed before in nature (Maupertuis, 1745, pp. 140–141).

The French encyclopedist Denis Diderot (1713–1784) also considered the possibility that living organisms result from natural processes involving selection in his *Lettre sur Les Aveugles, a l'Usage de Ceux qui Voyent . . .* (1749). Diderot (Paper 6) imagined a death-bed dialogue between the blind English mathematician Nicholas Sanderson and a Reverend Holmes in which the Blind Mr. Sanderson questions the arguments for the existence of God put forth by natural theologians. Mr. Sanderson raises the question as to whether Nature's variety can be explained without recourse to divine special creation. Could it possibly be that the living organisms we observe today are the outcome of natural processes operating in the past that involved the production of organisms by trial and error followed by the survival of the fit—"only those have remained whose physical equipment included no important contradiction and who were therefore able to subsist unaided and perpetuate themselves" (Paper 6, p. 22).

Diderot's ideas about the origin of modern-day species were more similar to those of Empedocles and Lucretius (Papers 1 and 2) than to the theories discussed by Maupertuis four years earlier in *Venus Physique* (Crocker, 1959; Bajema, 1982). The Lucretian possibility that Diderot summarized may have been responsible for Maupertuis publishing a similar statement one year later in his 1750 *Essay de Cosmologie* (Papers 7A and 7B). While both Maupertuis and Diderot discussed the possibility of gradual evolution in later publications (Maupertuis, 1751; Diderot, 1753), neither made any explicit statements about the possible role that selection might play in bringing about such gradual change (Glass, 1959; Crocker, 1959).

Georges Buffon, the great French naturalist who published the first of 44 volumes of his *Histoire Naturelle* in 1749, was impressed by the theories of Maupertuis (Glass, 1959; Woodward, 1960). Buffon remarked that "*Venus Physique* assembles more philosophical ideas than there are alltold in several great volumes on generation" and that Maupertuis "reasoned better than all those who have written before him on this subject" (Buffon, 1769, quoted in Glass, 1859). Unfortunately, Buffon did not construct a theory of evolution based on Maupertuis's ideas (Lovejoy, 1959; Glass, 1959), and he explicitly denied the possibility that any new species may have come into existence, despite infinite combinations of matter and of living things (Buffon, 1753, IV, pp. 182–183; see also Crocker, 1959, p. 128; Wilkie, 1956).

REFERENCES

Bajema, C. J., ed., 1982, *Artificial Selection and the Development of Evolutionary Theory,* Benchmark Papers in Systematic and Evolutionary Biology, vol. 4, Hutchinson Ross, Stroudsburg, Pa., 384p.

Buffon, G. comte de, 1753, *Histoire Naturelle Generale et Particuliere,* vol. IV, Paris, Imprimerie Royale, 540p.

Buffon, G. comte de, 1769, *Histoire Naturelle Generale et Particuliere,* Nouvelle Edition, vol. III, Paris.

Crocker, L., 1959, Diderot and Eighteenth Century Transformism, in *Forerunners of Darwin: 1745–1859,* B. Glass, O. Temkin, and W. Strauss, eds., Johns Hopkins University Press, Baltimore, Md., pp. 114–143.

Diderot, D., 1749, *Lettre sur les Aveugles, a l'Usage de Ceux qui Voyent . . . ,* Londres, 220p.

Diderot, D., 1753, *Pensees sur l'Interpretation de la Nature,* Paris, 206p.

Glass, B., 1959, Maupertuis, Pioneer of Genetics and Evolution, in *Forerunners of Darwin: 1745–1859,* B. Glass, O. Temkin, and W. Strauss, eds., Johns Hopkins University Press, Baltimore, Md., pp. 51–83.

Lovejoy, A., 1959, Buffon and the Problem of Species, in *Forerunners of Darwin: 1745–1859,* B. Glass, O. Temkin, and W. Strauss, eds., Johns Hopkins University Press, Baltimore, Md., pp. 84–113.

Maupertuis, P. L. M. de. (anonymous), 1745, *Venus Physique: Premiere Partie Contenant une Dissertation sur l'Origine des Hommes, et des*

Animaux; Seconde Partie Contenant une Dissertation sur l'Origine des Noirs, Paris, 194p.

Maupertuis, P. L. M. (Docteur Baumann), 1751, *Dissertatio Inauguralis de Universali Naturae Systemate, Pro Gradu Doctoris Habita,* Erlangen. (French translation, 1754, *Systeme de la Nature.*)

Wilkie, J., 1956, The Idea of Evolution in the Writings of Buffon, *Ann. Sci.* **12:**48-62, 212-227, 255-266.

Woodward, M., 1960, The Influence of the *Venus Physique* and *Systeme de la Nature* by P. L. M. Maupertuis on Eighteenth Century French Thought, M. A. thesis, University of Chicago.

6

LETTER ON THE BLIND, FOR THE USE OF THOSE WHO SEE

D. Diderot

[*Editor's Note:* In the original, material precedes and follows this excerpt.]

When he[1] was about to die, an extremely able minister of religion, the Reverend Gervaise Holmes, was summoned to his bedside. They conversed together about the existence of God, and I shall translate for you here the few fragments of their

[1] The blind English mathematician Nicholas Saunderson. The anecdote related here was invented by Diderot himself and caused considerable resentment in the Royal Academy.

conversation that have come down to us, for they are certainly worthy of our attention. The minister began by using the marvels of nature as an argument for God's existence.

"Oh, sir," the blind philosopher replied, "let us not talk of that great and beautiful spectacle which was never made for me! I was condemned to pass my life in darkness, and you are citing such marvels to me as I can never understand. For these things are proofs only to those who can see, like you. If you desire me to believe in God, then you must make me touch and feel him."

"Sir," the minister adroitly continued, "place your hands upon your own body and there you will sense the divinity that resides in the admirable mechanism of your own organs."

"Mr. Holmes," Saunderson replied, "I tell you again, nothing of all that can be as beautiful to me as it is to you. And even if the physical mechanism of animals is as perfect as you claim—and I am certainly willing to believe it is, for you are an honest man and incapable of any attempt to deceive me—what has that to do with a sovereignly intelligent being? If it is a matter of astonishment for you, then that may possibly be because you are in the habit of treating everything that is beyond your comprehension as a miracle. I myself have been an object of wonder to you so often that my opinion of what surprises you is inevitably a low one. People have been drawn to see me from all over England merely because they could not conceive how I was able to do geometry; you must admit that those people did not have very exact notions as to what is or is not possible. We encounter some phenomenon that is, in our opinion, beyond the powers of man, and immediately we say, 'It is the work of God.' Our vanity will admit no lesser explanation. Can we not reason with a little less pride and a little more philosophy? If nature presents us with a knot that is difficult to untie, then let us leave it as it is; let us not insist on cutting it there and then and on employing for the task the hand of a being who thereupon becomes a knot even more difficult to untie than the first. Ask an Indian why the globe remains suspended in the air and he will reply that it is borne on the back of an elephant. And on what does the elephant rest? On a tortoise. And the tortoise, who supports that? . . . This Indian fills you with pity, and yet one might answer, to you as to him: Mr. Holmes, my friend, confess your ignorance right away, and spare me the elephant and the tortoise."

Saunderson broke off for a moment; he seemed to be waiting for the minister to reply. But what method of attack could he use to convince a blind man? Mr. Holmes took advantage of the good opinion Saunderson had conceived of his own honesty and of the learning of Newton, Leibniz, Clarke,[2] and some others among his compatriots—the leading geniuses of the world, all of whom had been deeply impressed by the wonders of nature and recognized that they had been created by an intelligent being. This was, beyond a doubt, the strongest argument that the minister could present to Saunderson. And, indeed, the good blind man did admit that there would be some temerity in denying what a man such as Newton had not disdained to accept. Nevertheless, he went on to point out to the minister that Newton's evidence in this matter was not as strong for him, being blind, as that of nature in its entirety had been for Newton; and that whereas Newton's belief was based on the Word of God, he himself was reduced to taking the word of Newton.

"Just think, Mr. Holmes," he added, "to what extent I am forced to place my trust in your word and in Newton's. I cannot see anything, yet I am admitting that there is a wonderful order in the universe, though I hope that you will require no more of me. I am yielding to you on the present state of the universe so that you will permit me, in return, to think as I please about its former and first state, in regard to which you are no less blind than I. It is a matter in which you can produce no witnesses against me and in which your eyes are of no use to you whatever. You may imagine, then, if you wish, that the order you find so impressive has always existed. But please allow me to believe that such is not the case; that if we went back to the origin of things and time, if we were to sense matter as it first began to move and perceive chaos start to take some shape, then we should discover a whole multitude of monstrous, un-

[2] Samuel Clarke (1675–1729), English philosopher and theologian, author of a *Demonstration of the Existence and Attributes of God.*

formed shapes for every properly organized one. For though I cannot argue with you about the present state of things, I may at least question your beliefs on their former state. I may ask you, for example, who it was that told you or Leibniz or Clarke or Newton that during the short time when animals were first formed, some were not without a head and others without feet? I am in a position to maintain that some had no stomachs and others no intestines; that certain others, to whom the possession of a stomach, a palate, and a set of teeth seemed to promise longer life, ceased to exist because of some defect in the heart or lungs; that such monsters all died out, one after the other; that all the defective combinations created by matter have now disappeared and that only those have remained whose physical equipment included no important contradiction and who were therefore able to subsist unaided and perpetuate themselves.

"If we suppose all this, then if the first man had been born with a blocked larynx or had lacked suitable food or had possessed defective organs of generation or had not encountered a female or had combined with another species, then what, Mr. Holmes, would have become of the human race? It would have been included in the general purging of the universe; and this proud creature who calls himself man, dissolved and dispersed among the molecules of all matter, would have remained, perhaps forever, among the number of the possibles.

"If no ill-formed creatures had ever existed, then you would not fail to claim that there never will be any and that I am indulging in fantastic hypotheses; but the order of things is not so perfect," Saunderson went on, "that monstrous beings are not still produced from time to time." Then, turning his face to the minister, he added, "Look well at me, Mr. Holmes. I have no eyes. What did we do to God, you and I, that one of us should possess those organs and the other be deprived of them?"

Saunderson's air as he uttered these words was so sincere and so intense that the minister and the rest of the people present could not help but share his grief, and they began to weep bitterly over his fate. The blind man sensed this. "Mr. Holmes," he said, "the kindness of your heart was well known to me before, and I am very sensible of the proofs you are giving me of your goodness in these my last moments; but if I am dear to you, do not begrudge me upon my deathbed the consolation of never having been a cause of grief to anyone."

Then, in a firmer voice, he added, "I conjecture, then, that in the beginning, when fermenting matter was hatching out the universe, men such as myself were quite common. And why should I not assert what I believe about animals to be true of worlds also? How many maimed and botched worlds have melted away, re-formed themselves, and are perhaps dispersing again at any given moment, far away in space, where I cannot touch them and you cannot see them, but where motion still goes on, and will always go on, combining conglomerations of matter until they have attained a state of organization that will permit them to endure? Oh, philosophers, transport yourselves with me to the borders of the universe, beyond the point where I can still touch or you can still see organized entities; wander across the face of this new ocean and seek amidst its uncoordinated motions for some vestiges of that intelligent being whose wisdom you so much admire here on earth!

"But what need is there of taking you out of your own element? What is this world, Mr. Holmes? A composite entity subject to revolutions, all of which display a continuous tendency toward destruction; a rapid succession of beings that follow one another, thrust one another aside, then disappear; a passing symmetry; a momentary order. A short while ago, I was blaming you for judging the perfection of things according to the measure of your own abilities; I could accuse you now of measuring their duration according to the measure of your own lives. You are judging the discontinuous existence of the world as the ephemeral mayfly judges ours. The world is eternal for you in just the same way as you are eternal for the being that lives only an instant. Yet even so, the insect is more reasonable than you. Think of the tremendous succession of mayflies, generation after generation, that attests to your eternity! What an immense tradition! Yet we shall all pass away one day, and without any possibility of measuring either the real space that we once occupied or the exact time that we have lived. Time, space, and matter are perhaps only a point."

(24)

Saunderson grew somewhat more agitated during this discussion than his condition would permit of. He was overcome by an attack of delirium that lasted for several hours and from which he emerged at last only to cry, "*O God of Clarke and Newton, pity me now!*" and to die.

Such was Saunderson's end.

7A

Reprinted from pages 15–17 of *Essay de Cosmologie,* 1750

AVANT PROPOS OU L'ON EXAMINE LES PREUVES DE L'EXISTENCE DE DIEU, TIREES DES MERVEILLES DE LA NATURE

P. L. M. Maupertuis

[*Editor's Note:* In the original, material precedes and follows this excerpt. An English translation of this excerpt follows.]

Mais ne pourroit-on pas dire, que dans la combinaifon fortuite des productions de la Nature,

This facsimile was obtained through the courtesy of the History of Science Collections, Cornell University Libraries.

re, comme il n'y avoit que celles où ſe trouvoient certains rapports de convenance, qui puſſent ſubſiſter, il n'eſt pas merveilleux que cette convenance ſe trouve dans toutes les eſpèces qui actuellement exiſtent ? Le hazard, diroit-on, avoit produit une multitude innombrable d'Individus ; un petit nombre ſe trouvoit conſtruit de manière que les parties de l'Animal pouvoient ſatisfaire à ſes beſoins ; dans un autre infiniment plus grand, il n'y avoit ni convenance, ni ordre : tous ces derniers ont péri : des Animaux ſans bouche ne pouvoient pas vivre, d'autres qui manquoient

quoient d'organes pour la génération ne pouvoient pas ſe perpétuer ; les ſeuls qui ſoyent reſtés, ſont ceux où ſe trouvoient l'ordre & la convenance : & ces eſpèces que nous voyons aujourd'hui, ne ſont que la plus petite partie de ce qu'un deſtin aveugle avoit produit.

7B

Reprinted from *Pop. Sci. Monthly* **65**:242–243 (1904)

SOME EIGHTEENTH CENTURY EVOLUTIONISTS

A. Lovejoy

[*Editor's Note:* In the original, material precedes and follows this excerpt.]

Many, says Maupertuis, have found an evidence of design in the marvelous adaptation of the organs of animals to their needs. But "may we not say that, in the fortuitous combination of the productions of Nature, since only those creatures *could* survive in whose organization a certain degree of adaptation was present (*ou se trouvaient certains rapports de convenance*), there is nothing extraordinary in the fact that such adaptation is actually found in all those species which now exist? Chance, one might say, turned out a vast number of individuals; a small proportion of these were organized in such a manner that the animals' organs could satisfy their needs. A much greater number showed neither adaptation nor order; these last have all perished. . . . Thus the species which we see to-day are but a small part of all those that a blind destiny has produced."

Part IV

BRITISH THOUGHT ON NATURAL SELECTION THEORY PRIOR TO CHARLES DARWIN

Editor's Comments on Papers 8 Through 11

Erasmus Darwin (1731-1802), Charles Darwin's grandfather, developed a version of the theory that the inheritance of acquired characteristics brings about evolution. This theory was published in the first volume of his book *Zoonomia* (Darwin, 1794) six years before 1800, the year in which the great French scientist Jean Lamarck first championed his own theory that evolution occurred as the result of two processes: the "march of nature" due to the "power of progressively complicating organization" and the inheritance of characteristics acquired while individual organisms coped with the constraining influences of particular environmental factors (Lamarck, 1800). Erasmus Darwin never developed nor even discussed the theory that evolution might occur by selection working on randomly produced variations.

The first entry that Charles Darwin made in his first *Notebook on the Transmutation of Species* (de Beer, 1960-1967) was a paraphrase of a paragraph in his grandfather Erasmus's chapter on generation in *Zoonomia.* Charles took two of his grandfather's arguments seriously—that sexual reproduction, but not asexual reproduction, leads to the production of offspring that vary from their parents and that sexual reproduction is the culminating event in the life cycle and therefore is the "highest office in organization" (Kohn, 1980, pp. 83-87)

Erasmus Darwin did emphasize two very important dimensions of nature that Charles Darwin was to use in constructing his comprehensive theory as to the ecological causes and genetic consequences of selection. First, Erasmus Darwin championed the theory that nature was "one great slaughterhouse the warring world" rather than that nature was in a harmonious balance (see especially the excerpt from *The Temple of Nature* Darwin, 1803 presented as Paper 8). His grandfather's contention that nature was a Hobbesian "war of all against all" may have made young Charles Darwin more sympathetic to the "struggle for existence" view of nature that he read in Charles Lyell's *Principles of Geology* (1830-1833; see also Paper 12).

Second, Erasmus Darwin drew attention to the existence of adaptations that could not be explained in terms of benefit to the species. They could only be explained in terms of the reproductive advantage gained by the individual possessing the adaptation (Darwin, 1794, p. 503). Sometime before 1842, when he finally wrote out the first sketch of his species theory, Charles Darwin divided selection into natural and sexual selection and used sexual selection to explain the evolution of such adaptations (de Beer, 1958, pp. 48-49)

The English physician-anthropologist James Prichard (1786-1848) published his ideas on heredity and selection in his *Researches Into the Physical History of Mankind* in 1813. The second edition of Prichard's book appeared in 1826 and was read by Charles Lyell. Prichard's comments on variation and selective breeding in domesticated species (Prichard, 1826, II, pp. 557-558) probably served as the inspiration for Lyell's comments on the same topic in *Principles of Geology* (see Paper 12), comments that caused Charles Darwin to begin studying variation and selective breeding in domesticated species.

At an 1813 meeting of the Royal Society, William Wells (1757-1817), an American born English physician, used observations he made on a piebald woman as an opportunity to propose the theory that human racial differences in form and color were produced by selection. The paper presented at that meeting, "An Account of a Female of the White Race of Mankind, Part of Whose Skin Resembles that of a Negro; with Some Observations on the Causes of the Differences in Colour and Form Between the White and Negro Races of Men" (Paper 9), was reprinted posthumously in 1818 as an appendix to a new edition of his book *Two Essays: One on Dew and the Other on Single Vision With Two Eyes* (Wells, 1818).

Wells contended that climate and disease generated selection that operated on accidentally produced varieties of man, favoring those individuals "who would be better than the others to bear the diseases of the country" (Paper 9, p. 435). He contended that the disease-resistant varieties of humans "would consequently multiply, while the others would decrease, not only from their inability to sustain the attacks of disease, but from their incapacity of contending with their more vigorous neighbors" (Paper 9, pp. 435–436). While Wells drew an analogy between man's selective breeding of domesticated races and the evolution of human races (Paper 9, p. 435), he made no attempt to consider the possibility that selection of accidentally produced varieties might account for the origin of species in the rest of nature. An excellent account of the life of Wells and the impact that his ideas on the origin of human races had on his contemporaries as well as on Charles Darwin and Alfred Wallace, both of whom became aware of Wells's paper on human racial differences in 1865, has been published by Kentwood Wells (1973).

In 1831 Patrick Matthew (1790–1874), a Scottish botanical writer, published his views on the role that selection plays in nature in his book *Naval Timber and Arboriculture.* In this treatise on practical methods of raising trees for use in the construction of Royal Navy ships (Paper 10, pp. 106–108) Matthew contended that foresters, by their harvest methods, had been generating selection favoring the small growing and weakly varieties of each tree species. He also concluded that:

> Man's interference, by preventing this natural process of selection among plants, independent of the wider range of circumstances to which he introduces them, has increased the difference in varieties, particularly in the more domesticated kinds. (Paper 10, p. 308).

Matthew summarized his views on the role that a "natural process of selection" plays in modifying species in nature in an Appendix to his book (Paper 10, pp. 364–375, 381–388).

Matthew believed in geological catastrophism—a series of stable worlds interrupted by a series of geological catastrophies that destroyed most life on the earth. The survivors would then be exposed to radically different conditions and, having a largely unoccupied field for diversification, would be adapted by selection to new circumstances (Wells, 1973, p. 47). He also contended that the present assemblage of species had remained essentially unchanged "during the last forty centuries" (Paper 10, p. 381).

Matthew did not perceive natural selection as a mechanism leading to extinction of species via interspecific competition. He discussed the "process of selection in nature" not as a theory but as an axiom, a law of nature. In a May 12, 1860 letter to the *Gardener's Chronicle,* Matthew described how he discovered "the natural process of selection":

> To me the conception of this law of Nature came intuitively as a self-evident fact, almost without an effort of concentrated thought.

> Mr. Darwin here seems to have more merit in the discovery than I have had—to me it did not appear a discovery. He seems to have worked it out by inductive reason, slowly and with due caution to have made his way synthetically from fact to fact onwards; while with me it was by a general glance at the scheme of Nature that I estimated this select production of species as an a priori recognisable fact—an axiom, requiring only to be pointed out to be admitted by unprejudiced minds of sufficient grasp. (Matthew, 1860, p. 433)

Passages in Matthew's publications indicate that he was familiar with Malthusian population theory (Malthus, 1826). For example, he wrote of individuals "pressing on the means of subsistence" (Paper 10, p. 365). In contrast to the "total destruction and new creation" championed by the catastrophists of his day, Matthew contended that "There is more beauty and unity of design in this continual balancing of life to circumstance" (Paper 10, p. 384). The historian Kentwood Wells concluded that Matthew introduced his ideas on "the natural process of selection" in both his 1831 *Naval Timber and Arboriculture* and his 1839 *Emigration Fields* "not as theories to be discussed for their own sake, but as illustrations of the laws of nature to which men could look for guidance in the conduct of their own affairs" (Wells, 1973, p. 244). There is no evidence that Matthew's published thoughts on selection influenced Edward Blyth, Charles Darwin, or Alfred Wallace.

In 1835 the English naturalist Edward Blyth (1810–1873) pointed out that the struggle for existence produced selection in nature (Paper 11) but the only role he assigned to selection operating in nature was that of a conservative force maintaining the type. Blyth drew an analogy between man's selective breeding of domesticated species and the selection produced by the struggle for existence in the rest of nature: "The law which was intended by Providence to keep up the typical qualities of a species can be easily converted by man into a means of raising different varieties" (Paper 11, p. 46). He also gave an example of what Charles Darwin was to later classify as sexual selection. Thus all three forms of selection that Darwin was to later champion as the causes of "descent with modification" were discussed by Edward Blyth.

Blyth returned to the subject of adaptation by selection two years later (Blyth, 1837) and again restricted his discussion to selection operating in a conservative fashion. He did, however, go so far as to raise the question "May not, then, a large proportion of what are considered species have descended from a common parentage?" (Blyth, 1837, p. 147). The historian Kentwood Wells (1971) has concluded that William Lawrence's 1819 *Lectures on the Physiology, Zoology, and Natural History of Man* was a major source of Blyth's ideas on heredity and selection.

The anthropologist Loren Eiseley (1959, 1965, 1979) has contended that Charles Darwin may have learned about the importance of selection in nature by reading Blyth's papers. The entries that Charles Darwin made in his *Transmutation of Species* notebooks concerning Edward Blyth's 1835 and 1837 papers refer to the natural history of the wildlife Blyth discussed and to Blyth's ideas about instinct and reason, but not to Blyth's ideas about selection as a conservative force in nature (Schwartz, 1974). Darwin most likely did not record Blyth's ideas about selection because he was already aware of the struggle for existence (Paley, 1802; Darwin, Paper 8; Lyell, Paper 12, 1833) and of natural selection as a conservative force (Lyell, Paper 12, pp. 55-56). Charles Darwin's autobiographical description of the intellectual path that he took on his way to discovering the importance of selection in bringing about adaptive change is supported by the entries made in his *Transmutation of Species* notebooks. Darwin was looking for a process that changed species, not a process that merely maintained a species.

Blyth left England in 1841 for health reasons and spent the rest of his life in India. The Edward Blyth-Charles Darwin correspondence between 1854 and 1860 provides additional evidence that Blyth did not view selection as an evolutionary force (Beddall, 1971). He subscribed to the idea of multiple centers of creation. Blyth did write Darwin in 1855 asking whether Alfred Russel Wallace's newly published paper "On the Law That Regulates the Introduction of New Species" (Wallace, 1855) had unsettled Darwin's ideas concerning the persistence of species, but he did not go the next step in his letter and suggest that selection was the process by which new species were produced.

REFERENCES

Bajema, C. J., ed., 1982, *Artificial Selection and the Development of Evolutionary Theory,* Benchmark Papers in Systematic and Evolutionary Biology, vol. 4, Hutchinson Ross, Stroudsburg, Pa., 384p.

Beddall, B. G., 1971, "Notes for Mr. Darwin": Letters to Charles Darwin from Edward Blyth at Calcutta: A Study in the Process of Discovery, *J. Hist. Biol.* **6:**69-95.

Blyth, E., 1837, On the Psychological Distinctions Between Man and All Other Animals; and the Consequent Diversity of Human Influence Over the Inferior Ranks of Creation, from Any Mutual and Reciprocal Influence Excerised Among the Latter, *Mag. Nat. Hist.,* n.s., **1:**1-9, 77-85, 131-141.

Darwin, E., 1794, *Zoonomia; or the Laws of Organic Life,* vol. 1, J. Johnson, London, 586p.

Darwin, E., 1803, *The Temple of Nature; or, the Origin of Society: A Poem with Philosophical Notes,* J. Johnson, London, 124p.

de Beer, G., ed., 1958, *Evolution by Selection,* Cambridge University Press, Cambridge, 288p.

de Beer, G., ed., 1960-1967, *Darwin's Notebooks on Transmutation of Species.*
Part I. First Notebook (July 1837-February 1838) [Notebook B], *Br. Mus. (Nat. Hist.) Bull. Hist. Ser.* **2**(2):23-73 (1960).
Part II. Second Notebook (February 1838-July 1838) [Notebook C], *Br. Mus. (Nat. Hist.) Bull. Hist. Ser.* **2**(3):75-118 (1960).
Part III. Third Notebook (July 15, 1838-October 2, 1838) [Notebook D], *Br. Mus. (Nat. Hist.) Bull. Hist. Ser.* **2**(4):119-150 (1960).
Part IV. Fourth Notebook (October 1838-July 10, 1839) [Notebook E], *Br. Mus. (Nat. Hist.) Bull. Hist. Ser.* **2**(5):151-183 (1960).
Part V. Addenda and Corrigenda, *Br. Mus. (Nat. Hist.) Bull. Hist. Ser.* **2**(6):185-200 (1961).
Part V. Pages Excised by Darwin, *Br. Mus. (Nat. Hist.) Bull. Hist. Ser.* **3**(5):129-176 (1967).

Eiseley, L. C., 1959, Charles Darwin, Edward Blyth, and the Theory of Natural Selection, *Am. Philos. Soc. Proc.* **103:**94-114.

Eiseley, L. C., 1965, Darwin, Coleridge, and the Theory of Unconscious Creation, *Daedalus* **94:**588-602.

Eiseley, L. C., 1979, *Darwin and the Mysterious Mr. X: New Light on the Evolutionists,* Dutton, New York, 278p.

Kohn, D., 1980, Theories to Work By: Rejected Theories, Reproduction, and Darwin's Path to Natural Selection, *Studies Hist. Biol.* **4:**67-170.

Lamarck, J., 1800, *Discours d'Ouverture.* (Reprinted in 1801 as part of *System des Animaux sans Vertebres . . . ,* Chez l'Auteur, Paris, 432p.)

Lawrence, W., 1819, *Lectures on Comparative Anatomy, Physiology, Zoology, and the Natural History of Man,* J. Callow, London, 579p. (Pages 448-451 from this publication appear as Paper 8 in *Artificial Selection and the Development of Evolutionary Theory,* C. J. Bajema, ed., Benchmark Papers in Systematic and Evolutionary Biology, vol. 4, Hutchinson Ross, Stroudsburg, Pa., 1982, pp. 60-63.)

Lyell, C., 1830-1833, *Principles of Geology, Being an Attempt to Explain the Former Changes of the Earth's Surface, by Reference to Causes Now in Operation,* 3 vols., Murray, London, 511p., 330p., 398p. + 109p.

Malthus, T. R., 1826, *An Essay on the Principle of Population; Or, A View of its Past and Present Effects on Human Happiness; with an Inquiry Into Our Prospects Respecting the Future Removal or Mitigation of the Evils Which It Occasions,* sixth edition, 2 vols., Murray, London, 535p., 528p.

Matthew, P., 1831, *On Naval Timber and Arboriculture; with Critical Notes on Authors Who Have Recently Treated the Subject of Planting,* A. Black, Edinburgh, 391p.

Matthew, P., 1839, *Emigration Fields. North America, the Cape, Australia, and New Zealand, Describing These Countries, and Giving a Comparative View of the Advantages They Present to British Settlers,* Longman, London, 237p.

Matthew, P., 1860, Letter to Editor, *Gardener's Chronicle,* May 12, p. 433.

Paley, W., 1802, *Natural Theology: or, Evidences of the Existence and Attributes of the Deity, Collected from the Appearances of Nature,* Faulder, London, 586p.

Prichard, J. C., 1813, 1826, *Researches Into the Physical History of Man,* Arch, London. (Pages 557-558 of the 1826 edition appear as Paper 9 in *Artificial Selection and the Development of Evolutionary Theory,* C. J. Bajema, ed., Benchmark Papers in Systematic and Evolutionary Biology, vol. 4, Hutchinson Ross, Stroudsburg, Pa., 1982, pp. 75-76.)

Schwartz, J. S., 1974, Charles Darwin's Debt to Malthus and Edward Blyth, *J. Hist. Biol.* **7:**301-318.

Wallace, A. R., 1855, On the Law Which has Regulated the Introduction of New Species, *Ann. Mag. Nat. Hist.* **16:**184-196.

Wells, K. D., 1971, Sir William Lawrence (1783-1867): A Study of Pre-Darwinian Ideas on Heredity and Variation, *J. Hist. Biol.* **4:**319-361.

Wells, K. D., 1973, The Historical Context of Natural Selection: The Case of Patrick Matthew, *J. Hist. Biol.* **6:**225-258.

Wells, W. C., 1818, *Two Essays; One Upon Single Vision With Two Eyes; The Other on Dew . . . and An Account of a Female of the White Race of Mankind, Part of Whose Skin Resembles That of a Negro; With Some Observations on the Causes of the Differences in Colour and Forms Between the White and Negro Races of Men,* Black, Edinburgh, 439p.

8

Reprinted from pages 129–135 of *The Temple of Nature; or, the Origin of Society, A Poem with Philosophical Notes*, Johnson, London, 1803

OF GOOD AND EVIL.

E. Darwin

I. " HOW FEW," the MUSE in plaintive accents cries,
And mingles with her words pathetic sighs.—
" How few, alas! in Nature's wide domains
The sacred charm of SYMPATHY restrains!
Uncheck'd desires from appetite commence,
And pure reflection yields to selfish sense!
—Blest is the Sage, who learn'd in Nature's laws
With nice distinction marks effect and cause;
Who views the insatiate Grave with eye sedate,
Nor fears thy voice, inexorable Fate!

Blest is the Sage, l. 7.

Felix, qui potuit rerum cognoscere causas;
Quique metus omnes, et inexorabile fatum,
Subjecit pedibus, strepitumque Acherontis avari.
VIRG. Georg. II. 490.

"When War, the Demon, lifts his banner high,
And loud artillery rends the affrighted sky;
Swords clash with swords, on horses horses rush,
Man tramples man, and nations nations crush;
Death his vast sithe with sweep enormous wields,
And shuddering Pity quits the sanguine fields.

"The wolf, escorted by his milk-drawn dam,
Unknown to mercy, tears the guiltless lamb;
The towering eagle, darting from above,
Unfeeling rends the inoffensive dove;
The lamb and dove on living nature feed,
Crop the young herb, or crush the embryon seed.
Nor spares the loud owl in her dusky flight,
Smit with sweet notes, the minstrel of the night;
Nor spares, enamour'd of his radiant form,
The hungry nightingale the glowing worm;

The towering eagle, l. 19.
Torva leæna lupum sequitur, lupus ipse capellam,
Florentem cytisum sequitur lasciva capella. Virg.

Who with bright lamp alarms the midnight hour,
Climbs the green stem, and slays the sleeping flower.

"Fell Oestrus buries in her rapid course
Her countless brood in stag, or bull, or horse;
Whose hungry larva eats its living way,
Hatch'd by the warmth, and issues into day.
The wing'd Ichneumon for her embryon young
Gores with sharp horn the caterpillar throng.

Fell Oestrus buries, l. 29. The gadfly, bot-fly, or sheep-fly: the larva lives in the bodies of cattle throughout the whole winter; it is extracted from their backs by an African bird called Buphaga. Adhering to the anus it artfully introduces itself into the intestines of horses, and becomes so numerous in their stomachs, as sometimes to destroy them; it climbs into the nostrils of sheep and calves, and producing a nest of young in a transparent hydatide in the frontal sinus, occasions the vertigo or turn of those animals. In Lapland it so attacks the rein deer that the natives annually travel with the herds from the woods to the mountains. Lin. Syst. Nat.

The wing'd Ichneumon, l. 33. Linneus describes seventy-seven species of the ichneumon fly, some of which have a sting as long and some twice as long as their bodies. Many of them insert their eggs into various caterpillars, which when they are hatched seem for a time to prey on the reservoir of silk in the backs of those animals designed for their own use to spin a cord to support them, or a bag to contain them, while they change from their larva form to a butterfly; as I have seen in above fifty cabbage-caterpillars. The ichneumon larva then makes its way out of the caterpillar, and spins itself a small

The cruel larva mines its silky course,
And tears the vitals of its fostering nurse.
While fierce Libellula with jaws of steel
Ingulfs an insect-province at a meal;
Contending bee-swarms rise on rustling wings,
And slay their thousands with envenom'd stings.

"Yes! smiling Flora drives her armed car
Through the thick ranks of vegetable war;

cocoon like a silk worm; these cocoons are about the size of a small pin's head, and I have seen about ten of them on each cabbage caterpillar, which soon dies after their exclusion.

Other species of ichneumon insert their eggs into the aphis, and into the larva of the aphidivorous fly: others into the bedeguar of rose trees, and the gall-nuts of oaks; whence those excrescences seem to be produced, as well as the hydatides in the frontal sinus of sheep and calves by the stimulus of the larvæ deposited in them.

While fierce Libellula, l. 37. The Libellula or Dragon-fly is said to be a most voracious animal; Linneus says in their perfect state they are the hawks to naked winged flies; in their larva state they run beneath the water, and are the cruel crocodiles of aquatic insects. Syst. Nat.

Contending bee-swarms, l. 39. Stronger bee-swarms frequently attack weak hives, and in two or three days destroy them and carry away their honey; this I once prevented by removing the attacked hive after the first day's battle to a distinct part of the garden. See Phytologia, Sect. XIV. 3. 7.

Herb, shrub, and tree, with strong emotions rise
For light and air, and battle in the skies;
Whose roots diverging with opposing toil
Contend below for moisture and for soil;
Round the tall Elm the flattering Ivies bend,
And strangle, as they clasp, their struggling friend;
Envenom'd dews from Mancinella flow,
And scald with caustic touch the tribes below;
Dense shadowy leaves on stems aspiring borne
With blight and mildew thin the realms of corn;
And insect hordes with restless tooth devour
The unfolded bud, and pierce the ravell'd flower.

"In ocean's pearly haunts, the waves beneath
Sits the grim monarch of insatiate Death;
The shark rapacious with descending blow
Darts on the scaly brood, that swims below;

The shark rapacious, l. 57. The shark has three rows of sharp teeth within each other, which he can bend downwards internally to admit larger prey, and raise to prevent its return; his snout hangs

The crawling crocodiles, beneath that move,
Arrest with rising jaw the tribes above;
With monstrous gape sepulchral whales devour
Shoals at a gulp, a million in an hour.
—Air, earth, and ocean, to astonish'd day
One scene of blood, one mighty tomb display!
From Hunger's arm the shafts of Death are hurl'd,
And one great Slaughter-house the warring world!

so far over his mouth, that he is necessitated to turn upon his back, when he takes fish that swim over him, and hence seems peculiarly formed to catch those that swim under him.

The crawling crocodiles, l. 59. As this animal lives chiefly at the bottom of the rivers, which he frequents, he has the power of opening the upper jaw as well as the under one, and thus with greater facility catches the fish or water-fowl which swim over him.

One great slaughter-house, l. 66. As vegetables are an inferior order of animals fixed to the soil; and as the locomotive animals prey upon them, or upon each other; the world may indeed be said to be one great slaughter-house. As the digested food of vegetables consists principally of sugar, and from this is produced again their mucilage, starch, and oil, and since animals are sustained by these vegetable productions, it would seem that the sugar-making process carried on in vegetable vessels was the great source of life to all organized beings. And that if our improved chemistry should ever discover the art of making sugar from fossile or aerial matter without the assistance of vegetation, food for animals would then become as plentiful as water, and they might live upon the earth without preying on each other, as

[*Editor's Note:* Material has been omitted at this point.]

thick as blades of grass, with no restraint to their numbers but the want of local room.

It would seem that roots fixed in the earth and leaves innumerable waving in the air were necessary for the decomposition of water and air, and the conversion of them into saccharine matter, which would have been not only cumberous but totally incompatible with the locomotion of animal bodies. For how could a man or quadruped have carried on his head or back a forest of leaves, or have had long branching lacteal or absorbent vessels terminating in the earth? Animals therefore subsist on vegetables; that is they take the matter so prepared, and have organs to prepare it further for the purposes of higher animation and greater sensibility.

9

Reprinted from pages 425–439 of *Two Essays; One Upon Single Vision with Two Eyes; The Other on Dew . . . and An Account of a Female of the White Race of Mankind, Part of Whose Skin Resembles That of a Negro; With Some Observations on the Causes of the Differences in Colour and Form Between the White and Negro Races of Men*, A. Constable, London, 1818

ACCOUNT

OF A FEMALE

OF THE WHITE RACE OF MANKIND,

PART OF WHOSE SKIN

RESEMBLES THAT OF A NEGRO, &c.

W. C. Wells

INSTANCES of the absence of the black colour, in the whole or part of the skin in persons of the negro race, are not very uncommon; but there is, I believe, no one upon record of an individual of the white race having any part of the body, covered with a skin similar to that of a negro. The following account, therefore, of such an instance, will, perhaps, be acceptable to the philosophical public. I have been enabled to form it by the permission of Dr. Turner, one of my colleagues at St. Thomas's Hospital, into which the person, whose case I am to describe, was lately admitted by him, on account of some bodily ailment.

Hannah West, the subject of this account, was born in a village in Sussex, about three miles distant from the sea, and is now in the twenty-third year of her age. Both of her parents were natives of the same county. Her father was a footman in a gentleman's family, and died while she was very young. She cannot, therefore, remember his appearance; but she has never heard, that it was in any way extraordinary. Her mother is still alive, and has black hair and hazel eyes, but a fair skin, without any stain or mark upon it. West was the only child of her father; but her mother, having married a second time, has had eleven other children. Nine of these are living, all of whom are without any blackness of the skin. Her mother, she says, received a fright, while pregnant with her, by accidentally treading on a live lobster; and to this was attributed the blackness of part of her skin, which was observed at her birth.

West is somewhat above the middle size, is rather of a full habit, and till she came to London from Sussex, which was about four months ago, always enjoyed very good health. The hair of her head is of a light brown colour, and is very soft; her eyes of a faint blue; her nose prominent and a little aquiline; her lips thin; the skin of the greater portion of the uncovered parts of her body very white; in short, her appearance is in every respect, except the one which has been mentioned, that of a very fair female of the white race of mankind.

The parts covered by the black skin are, the left shoulder, arm, fore-arm, and hand. All these parts, however, are not universally black; for on the outside of the fore-arm, a little below the elbow, a stripe of white skin commences, about two inches in breadth, and differing in no circumstance from the skin of the other arm, which, proceeding upwards, gradually bends under the arm, and at the arm-pit joins with the white skin of the trunk of the body. The black skin, wherever it is contiguous to the white, terminates rather abruptly, so that its boundary may always be distinctly traced.

The colour of the black skin is not every where uniformly dark. Thus, the skin of the back of the hand, and of the wrist, is marked by fine lines of a reddish black, which cross one another at right angles, while the small rectangular spaces bounded by these lines are entirely black. Part of the cuticle of the hand having been removed by exciting a blister, the reddish lines were found to be the summits of very thin folds of the true skin, which were raised above its general level, and were less thickly covered with the black *rete mucosum* than the more depressed parts. Their reddish colour was, no doubt, occasioned by the external air, as the skin of the other hand was red from that cause. All the other parts of the black skin are fully as dark, as I found on making the comparison, as the corresponding parts of a dark negro, and are much darker than those of many negroes. One part, indeed, of her skin is considerably darker than the corresponding part in any negro whom I have seen; for the palm of her hand and inside of her fingers are black, whereas these parts in a negro are only of a tawny hue.

A considerable part of the black skin is as smooth to the touch, as the skin of the white arm; but the cuticular lines in the black arm, appeared everywhere stronger to the sight, than similar lines in the arm of a black man, whose skin I examined at the same time. In the greater part, however, of West's black skin, those lines sink deeper beneath its general surface, than the lines of any other human skin that I have seen, which was not evidently diseased. These depressions are extremely narrow, and proceed chiefly in one direction, obliquely upwards and inwards from the outer part of the arm. On removing a small portion of the cuticle, they were found to be occasioned by the sinking down of that membrane between very narrow and slightly elevated folds of the true skin, nearly contiguous to one another, which held the direction mentioned.

A great part of the black shoulder exhibits a singular appearance; for, near to the back bone, the skin, over an extent of six inches in length and two in breadth, resembles a thick coat of pitch, or black paint, which by drying had split into a great number of small square portions. The fissures in the skin are about a line in depth. Mr. James Wilson, teacher of anatomy, and fellow of the Royal Society, who saw this person once along with me, pulled away a little of this black matter, upon which several narrow processes of the skin, perpendicular to the plane of the part, became visible.

Winslow says, that the cuticle of a negro is black, and that the contrary supposition arose from its tenuity and transparency, in like manner as a thin film of black horn appears almost colourless. I have found by my own observations, that this opinion of Winslow is just; and I found also, that the cuticle of West's black skin is likewise dark. I may add, that the nails of her black fingers are darker than those of the white, and darker also than those of a negro's hand.

Sir Everard Home, who likewise saw this person once along with me, thought that the black arm smelt more strongly than the white. I made the experiment immediately after him, and thought so too But on repeating it several times with more attention, I could perceive no difference. It seems to me, indeed, from a similar experiment made on the arm of a dark negro, whose appearance did not lead me to suppose, that he had been very careful with respect to the cleanliness of his person, either that all negroes do not possess a strong smell, or that this does not proceed from all parts of their skin, since I could perceive no difference between the odour of his arm, and that of the white arm of West.

On the black fore-arm are about a dozen small hard substances, the largest of which are of the size of a common pea. Some of them are very black; others are less black, and one or two are of a reddish black colour. I thought, at first, that they consisted of thickened cuticle, but I found afterwards, that they readily bled upon being punctured with a needle.

The upper and outer part of the black arm has a number of very black hairs growing from it, some of which are three quarters of an inch long. The inner part of the arm, which is equally black, is free from hairs.

The black arm is as firm to the touch, and as fleshy as the white; and according to the young woman's own report, there is no difference in their strength or feelings of any kind.

The last circumstance which I shall mention concerning her is, that no change has taken place within her remembrance, either in the degree or extent of the blackness of her skin.

Two inferences may, I think, be made from what has been related respecting Hannah West.

The first is, that the blackness of the skin in negroes is no proof of their forming a different species of men from the white race.

When a white man is much exposed to the action of the sun, his skin becomes more or less brown, and as the intensity of this colour, after equal degrees of exposure, is generally proportional to the heat of the climate, it has hence been supposed, that the colour of negroes is derived from a very great degree of the same cause. But this conclusion seems to me very faulty. For, setting aside that a white man, rendered brown by the sun's rays, begets as white children as those of another of the same race, the colour of whose skin had never been altered, it appears to me probable, from observations lately made on two negroes, that the action of the sun tends rather to diminish than augment the colour of their race. Both of those persons were born in European settlements, and had been accustomed to have their bodies clothed, yet, in both, the trunk, arms, and lower extremities, were considerably darker than the face, and in one, were somewhat darker than the hands. But admitting this observation to be of no force, still it must be granted, in consequence of what has been said upon the state of part of West's skin,—that great heat is not indispensably necessary to render the human colour black; which is the second conclusion to be drawn from the account which has been given of her.

On considering the difference of colour between Europeans and Africans, a view has occurred to me of this subject, which has not been given by any author, whose works have fallen into my hands. I shall, therefore, venture to mention it here, though at the hazard of its being thought rather fanciful than just.

There is no circumstance, perhaps, in which these two races differ so much, as in their capacity to bear, with impunity, the action of the causes of many diseases. The fatality to Europeans of the climate of the middle parts of Africa, which are, however, inhabited by negroes without injury to their health, is well known. Let it then be supposed, that any number of Europeans were to be sent to that country, and that they were to subsist themselves by their bodily labour; it seems certain, that the whole colony would soon become extinct. On the other hand, the greater liability of negroes in Europe to be attacked with fatal diseases is equally well established. If, therefore, a colony of the former race were brought to Europe, and forced to labour in the open air for their subsistence, many of them would quickly die, and the remainder, from their inability to make great bodily exertions in cold weather, and their being frequently diseased, would be prevented from working an equal number of days in the year with the whites. The consequence would be, that without taking farther into account the unfriendliness of the climate to them, their gains would be inadequate to the maintenance of themselves and their families. They would thence become feeble, and be rendered still more incapable of supporting life by their labour. In the mean time, their children would die from want, or diseases induced by deficient or improper nourishment, and in this way, a colony of the negro race in a cold country would quickly cease to exist.

This difference in the capacity of the two races to resist the operation of the causes of many diseases, I assume as a fact, though I am utterly unable to explain it. I do not, however, suppose, that their different susceptibility of diseases depends, properly, on their difference of colour. On the contrary, I think it probable, that this is only a sign of some difference in them, which, though strongly manifested by its effects in life, is yet too subtle to be discovered by an anatomist after death; in like manner as a human body, which is incapable of receiving the small-pox, differs in no observable thing from another, which is still liable to be affected with that disease.

Regarding then as certain, that the negro race are better fitted to resist the attacks of the diseases of hot climates than the white, it is reasonable to infer, that those, who only approach the black race, will be likewise better fitted to do so, than others who are entirely white. This is, in fact, found to be true, with regard to the mixture of the two races; since mulattoes are much more healthy in hot climates than whites. But amongst men, as well as among other animals, varieties of a greater or less magnitude are constantly occurring. In a civilized country, which has been long peopled, those varieties, for the most part, quickly disappear, from the intermarriages of different families. Thus, if a very tall man be produced, he very commonly marries a woman much less than himself, and their progeny scarcely differs in size from their countrymen. In districts, however, of very small extent, and having little intercourse with other countries, an accidental difference in the appearance of the inhabitants will often descend to their late posterity. The clan of the Macras, for instance, possess both sides of Loch-Duich in Scotland; but those who inhabit one side of the loch are called the black Macras, and the others the white, from a difference which has always been observed in their complexions. Again, those who attend to the improvement of domestic animals, when they find individuals possessing, in a greater degree than common, the qualities they desire, couple a male and female of these together, then take the best of their offspring as a new stock, and in this way proceed, till they approach as near the point in view, as the nature of things will permit. But, what is here done by art, seems to be done, with equal efficacy, though more slowly, by nature, in the formation of varieties of mankind, fitted for the country which they inhabit. Of the accidental varieties of man, which would occur among the first few and scattered inhabitants of the middle regions of Africa, some one would be better fitted than the others to bear the diseases of the country. This race would consequently multiply, while the others would decrease, not only from their inability to sustain the attacks of disease, but from their incapacity of contending with their more vigorous neighbours. The colour of this vigorous race I take for granted, from what has been already said, would be dark. But the same disposition to form varieties still existing, a darker and a darker race would in the course of time occur, and as the darkest would be the best fitted for the climate, this would at length become the most prevalent, if not the only race, in the particular country in which it had originated.

In like manner, that part of the original stock of the human race, which proceeded to the colder regions of the earth, would in process of time become white, if they were not originally so, from persons of this colour being better fitted to resist the diseases of such climates, than others of a dark skin.

The cause which I have stated, as likely to have influence on the colour of the human race, would necessarily operate chiefly during its infancy, when a few wandering savages, from ignorance and improvidence, must have found it difficult to subsist throughout the various seasons of the year, even in countries the most favourable to their health. But, when men have acquired the knowledge of agriculture, and other arts, and in consequence adopt a more refined mode of life, it has been found, that an adherence to their ancient customs and practices will preserve them long as a distinct race from the original inhabitants of the country to which they had emigrated. Examples of this kind are frequent in the islands in the eastern seas in the torrid zone, where the inhabitants of the sea-coast, evidently strangers, are in some degree polished, and of a brown colour, while the ancient natives, who live in the interior parts, are savage and black. Similar facts occur in respect to other species of animals. It seems certain, for instance, that fine woolled sheep, like the Spanish, never both arose and sustained their breed in the northern parts of Europe; yet, by care, this feeble race, after being formed in Spain, has been propagated and preserved in very cold countries. Thus the late Mr. Dryander, the learned librarian of the Royal Society, informed me, that the breed of fine woolled Spanish sheep had been kept perfect in Sweden during a very long term of years, I think he said a century. If, then, my memory be accurate upon this point, we have here an example of a variety of animals, much more liable to be affected by external circumstances than the human race, being preserved without change, in a country very different from their own, by assimilating their new state as much as possible to their old, during at least fifty generations, that is, during a period equivalent to 1500 years in the history of man.

Hitherto, while speaking of the external appearance of negroes, I have taken notice only of their colour. I shall now say a few words upon their woolly hair, and, according to our notions of beauty, the deformity of their features.

There are several facts which seem to show, that these circumstances are somehow connected with their low state of civilization.

First; the black inhabitants of the Indian Peninsula within the Ganges, who, compared with the African negroes, may be regarded as a polished people, have hair and features much less dissimilar to the European.

Secondly; Woolly heads, and deformed features, appear again, as we proceed further to the east, among the savage inhabitants of New Guinea, and the adjacent islands, at the distance nearly of half of the circumference of our globe from Africa, and consequently without the smallest probability of any communication having ever existed between the two countries.

Lastly; it appears probable from the reliques of ancient art, that the early inhabitants of Egypt were of the negro race. If, then, the negroes of Africa were ever to be civilized, their woolly hair and deformed features would, perhaps, in a long series of years, like those of the Egyptians, be changed. On the other hand, their present external appearance may possibly be regarded not only as a sign, but as a cause of their degraded condition, by preventing, in some unknown way, the proper developement of their mental faculties; for the African negroes have in all ages been slaves; and the negroes in the eastern seas are in no instance, I believe, masters of their handsomer neighbours, but are in many places in entire subjection to them, though the latter be frequently less numerous.

It will no doubt be objected to what I have advanced respecting the difference of colour between Europeans and Africans, that the Indian inhabitants of the greater part of the immense continent of America have skins nearly of one hue. Plausible reasons may, I think, be given for this fact, consistently with what has been said upon the colour of the two former races; but I forbear trespassing any longer upon the time of the reader, in discussing a subject which admits only of conjectural reasoning.

THE END.

10

Reprinted from pages 106–108, 306–308, 364–375, and 381–388 of *On Naval Timber and Arboriculture; with Critical Notes on Authors Who Have Recently Treated the Subject of Planting,* A. Black, Edinburgh, 1831

ON NAVAL TIMBER AND ARBORICULTURE; WITH CRITICAL NOTES ON AUTHORS WHO HAVE RECENTLY TREATED THE SUBJECT OF PLANTING

P. Matthew

NURSERIES.

Much of the luxuriance and size of timber depending upon the particular variety of the species, upon the treatment of the seed before sowing, and upon the treatment of the young plant, and as this fundamental subject is neither much attended to nor generally understood, we shall take it up *ab initio.*

The consequences are now being developed of our deplorable ignorance of, or inattention to, one of the most evident traits of natural history, that vegetables as well as animals are generally liable to an almost unlimited diversification, regulated by climate, soil, nourishment, and new commixture of already formed varieties. In those with which man is most intimate, and where his agency in throwing them from their natural locality and dispositions has brought out this power of diversification in stronger shades, it has been forced upon his notice, as in man himself, in the dog, horse, cow, sheep, poultry,—in the apple, pear, plum, gooseberry, potato, pea, which sport in infinite varieties, differing considerably in size, colour, taste, firmness of texture, period of growth, almost in every recognisable quality. In all these kinds man is influencial in preventing deterioration, by careful selection of the largest or most valuable as breeders; but in timber trees the opposite course has been pursued. The large growing varieties being so long of coming to produce seed, that many plantations are cut down before they reach this maturity, the small growing and weakly varieties, known by early and extreme seeding, have been continually selected as reproductive stock, from the ease and conveniency with which their seed could be procured; and the husks of several kinds of these invariably kiln-dried *, in order that the seeds might be the more easily extracted! May we, then, wonder that our plantations are occupied by a sickly short-lived puny race, incapable of supporting existence in situations where their own kind had formerly flourished—particularly evinced in the genus Pi-

* If the heat and evaporation of a gardener's pocket for several days be sufficient to render the seeds of melons and gourds productive of plants of earlier maturity, that is less disposed to extension and more to reproduction,—what may be expected from kiln-drying fir-cones?

nus, more particularly in the species Scots fir; so much inferior to those of Nature's own rearing, where only the stronger, more hardy, soil-suited varieties can struggle forward to maturity and reproduction?

We say that the rural economist should pay as much regard to the breed or particular variety of his forest trees, as he does to that of his live stock of horses, cows, and sheep. That nurserymen should attest the variety of their timber plants, sowing no seeds but those gathered from the largest, most healthy, and luxuriant growing trees, abstaining from the seed of the prematurely productive, and also from that of the very aged and over-mature; as they, from animal analogy, may be expected to give an infirm progeny, subject to premature decay.

[*Editor's Note:* Material has been omitted at this point.]

(306)

We would ask how our author is enabled to assume, as an axiom, that trees produce the best timber in their natural locality? We would also desire some *rational* information to shew in what manner pruning up can in any way conduce generally, to the increase of the timber, or to the enlargement of one-stemmed vegetables. A tree naturally rises in one stem. It throws out its branches in the disposition most favourable to draw the fullest benefit from the light and air. It of its own accord (that is when man does not meddle), gradually raises its pyramidal centre, with proportional lateral spread, as high as is befitting, for the fullest expansion of the individual, under the circumstances of its location. Man may mar this beautiful natural balance easier than decypher the proximate cause he may throw the new deposit of wood in greater proportion upon the upper part of the stem, rendering his beam more suitable from equality of thickness, and particularly in pines, of cleaner, smaller growthed, more durable timber, thence more valuable. But the tree will neither produce the same quantity of measurable timber in a considerable number of years, nor will it ultimately reach to nearly the same size, nor continue life nearly so long, as when left to itself. Man's interference is useful in removing competitors, in giving it lateral room for extension, in *training* it skilfully to one leader and subordinate equality of feeders, should transplanting, early pruning up, or other cause, destroy the natural regular pyramidal disposition—not in pruning it up, thus reducing it to narrower compass, and destroying its balance to the locality.

The use of the infinite seedling varieties in the families of plants, even in those in a state of nature, differing in luxuriance of growth and local adaptation, seems to be to give one individual (the strongest best circumstance-suited) superiority over others of its kind around, that it may, by overtopping and smothering them, procure room for full extension, and thus affording, at the same time, a continual selection of the strongest, best circumstance-suited, for reproduction. Man's interference, by preventing this natural process of selection among plants, independent of the wider range of circumstances to which he introduces them, has increased the difference in varieties, particularly in the more domesticated kinds; and even in man himself, the greater uniformity, and more general vigour among savage tribes, is referrible to nearly similar selecting law—the weaker individual sinking under the ill treatment of the stronger, or under the common hardship.

As our author's premises thus appear neither self-evident, nor supported by facts, it might seem unfair, at least it would be superfluous, to proceed to the consideration of his conclusions and corollaries.

[*Editor's Note:* Material has been omitted at this point.]

Note B.

There is a law universal in nature, tending to render every reproductive being the best possibly suited to its condition that its kind, or that organized matter, is susceptible of, which appears intended to model the physical and mental or instinctive powers, to their highest perfection, and to continue them so. This law sustains the lion in his strength, the hare in her swiftness, and the fox in his wiles. As Nature, in all her modifications of life, has a power of increase far beyond what is needed to supply the place of what falls by Time's decay, those individuals who possess not the requisite strength, swiftness, hardihood, or cunning, fall prematurely without reproducing—either a prey to their natural devourers, or sinking under disease, generally induced by want of nourishment, their place being occupied by the more perfect of their own kind, who are pressing on the means of subsistence. The law of entail, necessary to hereditary nobility, is an outrage on this law of nature which she will not pass unavenged—a law which has the most debasing influence upon the energies of a people, and will sooner or later lead to general subversion, more especially when the executive of a country remains for a considerable time efficient, and no effort is needed on the part of the nobility to protect their own, or no war to draw forth or preserve their powers by exertion. It is all very well, when, in stormy times, the baron has every faculty trained to its utmost ability in keeping his proud crest aloft. How far hereditary nobility, under effective government, has operated to retard "the march of intellect," and deteriorate the species in modern Europe, is an interesting and important question. We have seen it play its part in France; we see exhibition of its influence throughout the Iberian peninsula, to the utmost degradation of its victims. It has rendered the Italian peninsula, with its islands, a blank in the political map of Europe. Let the panegyrists of hereditary nobility, primogeniture, and entail, say what these countries might not have been but for the baneful influence of this unnatural custom. It is an eastern proverb, that no king is many removes from a shepherd. Most conquerors and founders of dynasties have followed the plough or the flock. Nobility, to be in the highest perfection, like the finer varieties of fruits, independent of having its vigour excited by regular married alliance with wilder stocks, would require stated complete renovation, by selection anew, from among the purest crab. In some places, this renovation would not be so soon requisite as in others, and, judging from facts, we would instance Britain as perhaps the soil where nobility will continue the longest untainted. As we advance nearer to the equator, renovation becomes sooner necessary, excepting at high elevation—in many places, every third generation, at least with the Caucasian breed, although the finest stocks be regularly imported. This renovation is required as well physically as morally.

It is chiefly in regard to the interval of time between the period of necessary feudal authority, and that when the body of the population having acquired the power of self-government from the spread of knowledge, claim a community of rights, that we have adverted to the use of war. The manufacturer, the merchant, the sailor, the capitalist, whose mind is not corrupted by the indolence induced under the law of entail, are too much occupied to require any stimulant beyond what the game in the wide field of commercial adventure affords. A great change in the circumstances of man is obviously at hand.

In the first step beyond the condition of the wandering savage, while the lower classes from ignorance remained as helpless children, mankind naturally fell into clans under paternal or feudal government; but as children, when grown up to maturity, with the necessity for protection, lose the subordination to parental authority, so the great mass of the present population requiring no guidance from a particular class of feudal lords, will not continue to tolerate any hereditary claims of authority of one portion of the population over their fellow-men; nor any laws to keep up rank and wealth corresponding to this exclusive power.—It would be *wisdom* in the noblesse of Europe to abolish every claim or law which serves to point them out a separate class, and, as quickly as possible, to merge themselves into the mass of the population. It is a law manifest in nature, that when the use of any thing is past, its existence is no longer kept up.

Although the necessity for the existence of feudal lords is past, yet the same does not hold in respect to a hereditary head or King; and the stability of this head of the government will, in no way, be lessened by such a change. In the present state of European society, perhaps no other rule can be so mild and efficient as that of a liberal benevolent monarch, assisted by a popular representative Parliament. The poorest man looks up to his king as his own, with affection and pride, and considers him a protector; while he only regards the antiquated feudal lord with contempt. The influence of a respected hereditary family, as head of a country, is also of great utility in forming a principle of union to the different members, and in giving unity and stability to the government.

In respect to our own great landholders themselves, we would ask, where is there that unnatural parent—that miserable victim of hereditary pride—who does not desire to see his domains equally divided among his own children? The high paid sinecures in church and state will not much longer be a great motive for keeping up a powerful family head, whose influence may burthen their fellow-citizens with the younger branches. Besides, when a portion of land is so large, that the owner cannot have an individual acquaintance and associations with every stream, and bush, and rock, and knoll, the deep enjoyment which the smaller native proprietor would have in the peculiar features, is not called forth, and is lost to man. The abolition of the law of entail and primogeniture, will, in the present state of civilization, not only add to the happiness of the proprietor, heighten morality, and give much greater stability to the social order, but will also give a general stimulus to industry and improvement, increasing the comforts and elevating the condition of the operative class.

In the new state of things which is near at hand, the proprietor and the mercantile class will amalgamise,—employment in useful occupations will not continue to be held in scorn,—the merchant and manufacturer will no longer be barely tolerated to exist, harassed at every turn by imposts and the interference of petty tyrants;—Government, instead of forming an engine of oppression, being simplified and based on morality and justice, will become a cheap and efficient protection to person and property; and the necessary taxation being levied from property alone, every individual will purchase in the cheapest market, and sell the produce of his industry in the dearest. This period might, perhaps, be accelerated throughout Europe, did the merchants and capitalists only know their own strength. Let them, as citizens of the world, hold annual congress in some central place, and deliberate on the interests of man, which is their own, and throw the whole of their influence to support liberal and just governments, and to repress slavery, crime, bigotry—tyranny in all shapes. A Rothschild might earn an unstained fame, as great as yet has been attained by man, by organizing such a power, and presiding at its councils.

Note C.

The influence of long continued impression, constituting instinct or habit of breed, is a curious phenomenon in the animal economy. Our population in the eastern maritime districts of Britain, descended principally from the Scandinavian rover, though devoted for a time to agricultural or mechanical occupation, betake themselves, when opportunity offers, to their old element, the ocean *,

* The habit of breed is apparent in many places of the world. Where a fine river washes the walls of some of the internal towns of France, scarce a boat is to be seen, except the long tract-boats employed in the

and launch out upon the " wintry wave" with much of the same home-felt composure as does the white polar bear. They roam over every sea and every shore, from Behring's Straits to Magellan's, with as little solicitude as the Kelt over his own misty hill, overcoming, in endurance, the native of the torrid zone under his vertical sun, and the native of the frigid among his polar snows.

To what may we ascribe the superiority of this portion of the Caucasian breed,—may it arise in part from its repeated change of place under favourable circumstances? Other races have migrated, but not like this, always as conqueror. The Jew has been a stroller in his time; but he has improved more in mental acumen and cunning—not so much in heroism and personal qualities: his proscribed condition will account for this. The Caucasian in its progress, will also have mingled slightly, and, judging from analogy, perhaps advantageously, with the finer portion of those whom it has overwhelmed. This breed, by its wide move across the Atlantic, does not seem at all to have lost vigour, and retains the nautical and roving instinct unimpaired, although the American climate is certainly inferior to the European. It is there rapidly moving west, and may soon have described one of the earth's circles. A change of seed, that is, a change of place, within certain limits of latitude, is well known

conveyance of fire-wood—nobody thinks of sailing for pleasure. The Esquimaux, and the Red Indian of North America, inhabiting the same country, shew an entirely distinct habit of breed. The Black and the Copper-coloured native of the Australian Islands, are equally opposed in instinctive habit.

to be indispensable to the more sturdy growth and health of many cultivated vegetables; it is probable that this also holds true of the human race. There are few countries where the old breed has not again and again sunk before the vigour of new immigration; we even see the worn out breed, chased from their homes to new location, return, after a time, superior to their former vanquishers, or gradually work their way back in peace, by superior subsisting power: this is visible in France, where the aboriginal sallow Kelt, distinguished by high satyr-like feature, deep-placed sparkling brown or grey eye, narrowed lower part of the face, short erect vertebral column, great mental acuteness, and restless vivacity, has emerged from the holes of the earth, the recesses of the forests and wastes, into which it had been swept before the more powerful blue-eyed Caucasian; and being a smaller, more easily subsisting animal, has, by starving and eating out, been gradually undermining the breed of its former conquerors. The changes which have been taking place in France, and which, in many places, leave now scarcely a trace of the fine race which existed twenty centuries ago, may, however, in part, be accounted for by the admixture of the Caucasian and Keltic tending more to the character of the latter, from the latter being a purer and more fixed variety, and nearer the original type or medium standard of man; and from the warm dry plains of France (much drier from cultivation and the reduction of the forests), having considerable influence to increase this bias: In some of the south-eastern departments, more immediately in the tide of the ingress of the Caucasian, where the purest current has latest flowed, and the climate is more suitable, and also in some of the maritime districts, where the air is moister, and to which they have been seaborn at a later period, the Caucasian character is still prominent. Something of this, yet not so general, is occurring in Britain, where the fair bright-blooded race is again giving place to the darker and more sallow. This may, however, be partly occasioned by more of artificial heat and shelter and other consequences of higher civilization. There seems to be something connected with confinement and sedentary life, with morbid action of the liver, or respiratory or transpiratory organs, which tend to this change under dry and hot, and especially confined atmosphere. Perhaps imagination is also a worker here; and the colour most regarded, as snow in cold countries, black among colliers, white among bleachers, or even the dark colour of dress, may produce its peculiar impression, and our much looked-up-to Calvinistic priesthood, from the pulpit, disseminate darkness as well as light.

Our own Kelt has indubitably improved much since, *par necessité,* he took to the mountain; but, though steadily enduring, when there is mental excitement, he has acquired a distaste to dull hopeless unceasing labour, and would fare scantily and lie hard, rather than submit to the monotonous industry of the city operative, or the toil of the agricultural drudge. Though once a fugitive, the Kelt is now, in moral courage and hardihood, equal perhaps to any other, yet he still trembles to put foot on ocean.

Notwithstanding that change of place, simply, may have impression to improve the species, yet is it more to circumstances connected with this change, to which the chief part of the improvement must be referred. In the agitation which accompanies emigration, the ablest in mind and body—the most powerful varieties of the race will be thrown into their natural position as leaders, impressing the stamp of their character on the people at large, and constituting the more reproductive part; while the feebler or more improvident varieties will generally sink under the incidental hardships. When a swarm emigrates from a prosperous hive, it also will generally consist of the more adventurous stirring spirits, who, with the right of conquerors, will appropriate the finest of the indigenæ which they overrun; their choice of these being regulated by personal qualities, not by the adventitious circumstances of wealth or high birth—a regard to which certainly tends to deteriorate the species, and is one of the causes which renders the noblesse of Europe comparatively inferior to the Asiatic, or rather the Christian noblesse to the Mahometan.

It has been remarked, that our finest, most acute population, exist in the neutral ground, where the Caucasian and Keltic have mixed, but this may arise from other causes than admixture. Our healthiest and poorest country borders the Highlands, and the population enjoy more of the open air. Our eastern population, north of the natural division of Flamboroughead, are also harder and sharper featured, and keener witted, than those southward, who may be styled our fen-bred. There is no doubt more of Keltic blood mingled with the north division; but the sea-born breeds have also been different, those more northerly being Scandinavian, and the more southerly consisting of the native of Lower Germany and the heavy Fleming. The placid-looking Englishman, more under the control of animal enjoyment, though perhaps not so readily acute, excels in the no less valuable qualities of constancy and bodily powers of exertion; and when properly taught under high division of labour, becomes a better operative in his particular employment, and even will sometimes extend scientific discovery further, than his more mercurial northern neighbour, who, from his quick wits being generally in advance of his manual practice, seldom attains to the dexterity which results from the combination of continued bodily action and restricted mental application. There exists, however, very considerable intellectual capacity in this English breed, but it too frequently is crushed under the preponderance of the animal part, affording that purest specimen of vulgarity, the English clown. But, independently of climate and breed, a great part of the low Englander's obtuseness is referable to his being entailed lord of the soil, under poor-rate law, contravening a natural law (see note B), so that, when unsuccessful or out of employment, he, without effort to obtain some new means of independent subsistence, sinks into the parish or work-house labourer. On the contrary, the Scotsman, with no resource but in himself, with famine always in the vista, as much in his view as a principle of action in material affairs as his strong perception of the right in moral, and also under the stimulus of a high pride, leaves no means untried at home; and, when fairly starved out of his native country, among various resources, often invades the territory of his more easy-minded southern neighbour, where his acuteness seldom fails to find out a convenient occupation, in which manual dexterity is second to economy and forethought—his success exciting the wonder and envy of the dull-witted native.

It would appear, that the finest portion, at least apparently so, of the north temperate zone, between the parallels of 30° and 48° latitude, when nearly of the level of the ocean, is not so favourable for human existence as the more northern part between 50° and 60°, or even the torrid zone. The native of the north of Europe has a superior development of person, and a much longer reproductory life than the native of the south, which more than counterbalances the earlier maturity of the latter in power of increase. Independent of the great current of population setting south in the northern part of the temperate zone, there seems even to be some tendency to a flux northward, from the confines of the torrid; but this arises rather from the unsteadiness of the seasons, and consequent deficit of food, at particular times, than from a steady increase of population.

[*Editor's Note:* Material has been omitted at this point.]

Throughout this volume, we have felt considerable inconvenience, from the adopted dogmatical classification of plants, and have all along been floundering between species and variety, which certainly under culture soften into each other. A particular conformity, each after its own kind, when in a state of nature, termed species, no doubt exists to a considerable degree. This conformity has existed during the last forty centuries. Geologists discover a like particular conformity—fossil species—through the deep deposition of each great epoch, but they also discover an almost complete difference to exist between the species or stamp of life, of one epoch from that of every other. We are therefore led to admit, either of a repeated miraculous creation; or of a power of change, under a change of circumstances, to belong to living organized matter, or rather to the congeries of inferior life, which appears to form superior. The derangements and changes in organized existence, induced by a change of circumstance from the interference of man, affording us proof of the plastic quality of superior life, and the likelihood that circumstances have been very different in the different epochs, though steady in each, tend strongly to heighten the probability of the latter theory.

When we view the immense calcareous and bituminous formations, principally from the waters and atmosphere, and consider the oxidations and depositions which have taken place, either gradually, or during some of the great convulsions, it appears at least probable, that the liquid elements containing life have varied considerably at different times in composition and in weight; that our atmosphere has contained a much greater proportion of carbonic acid or oxygen; and our waters, aided by excess of carbonic acid, and greater heat resulting from greater density of atmosphere, have contained a greater quantity of lime and other mineral solutions. Is the inference then unphilosophic, that living things which are proved to have a circumstance-suiting power—a very slight change of circumstance by culture inducing a corresponding change of character—may have gradually accommodated themselves to the variations of the elements containing them, and, without new creation, have presented the diverging changeable phenomena of past and present organized existence.

The destructive liquid currents, before which the hardest mountains have been swept and comminuted into gravel, sand, and mud, which intervened between and divided these epochs, probably extending over the whole surface of the globe, and destroying nearly all living things, must have reduced existence so much, that an unoccupied field would be formed for new diverging ramifications of life, which, from the connected sexual system of vegetables, and the natural instincts of animals to herd and combine with their own kind, would fall into specific groups, these remnants, in the course of time, moulding and accommodating their being anew to the change of circumstances, and to every possible means of subsistence, and the millions of ages of regularity which appear to have followed between the epochs, probably after this accommodation was completed, affording fossil deposit of regular specific character.

There are only two probable ways of change—the above, and the still wider deviation from present occurrence,—of indestructible or molecular life (which seems to resolve itself into powers of attraction and repulsion under mathematical figure and regulation, bearing a slight systematic similitude to the great aggregations of matter), gradually uniting and developing itself into new circumstance-suited living aggregates, without the presence of any mould or germ of former aggregates, but this scarcely differs from new creation, only it forms a portion of a continued scheme or system.

In endeavouring to trace, in the former way, the principle of these changes of fashion which have taken place in the domiciles of life, the following questions occur: Do they arise from admixture of species nearly allied producing intermediate species? Are they *the diverging ramifications* of the living principle under modification of circumstance? Or have they resulted from the combined agency of both? Is there only one living principle? Does organized existence, and perhaps all material existence, consist of one Proteus principle of life capable of gradual circumstance-suited modifications and aggregations, without bound under the solvent or motion-giving principle, heat or light? There is more beauty and unity of design in this continual balancing of life to circumstance, and greater conformity to those dispositions of nature which are manifest to us, than in total destruction and new creation. It is improbable that much of this diversification is owing to commixture of species nearly allied, all change by this appears very limited, and confined within the bounds of what is called Species; the progeny of the same parents, under great difference of circumstance, might, in several generations, even become distinct species, incapable of co-reproduction.

The self-regulating adaptive disposition of organized life may, in part, be traced to the extreme fecundity of Nature, who, as before stated, has, in all the varieties of her offspring, a prolific power much beyond (in many cases a thousandfold) what is necessary to fill up the vacancies caused by senile decay. As the field of existence is limited and pre-occupied, it is only the hardier, more robust, better suited to circumstance individuals, who are able to struggle forward to maturity, these inhabiting only the situations to which they have superior adaptation and greater power of occupancy than any other kind; the weaker, less circumstance-suited, being prematurely destroyed. This principle is in constant action, it regulates the colour, the figure, the capacities, and instincts; those individuals of each species, whose colour and covering are best suited to concealment or protection from enemies, or defence from vicissitude and inclemencies of climate, whose figure is best accommodated to health, strength, defence, and support; whose capacities and instincts can best regulate the physical energies to self-advantage according to circumstances—in such immense waste of primary and youthful life, *those* only come forward to maturity from the strict ordeal by which Nature tests their adaptation to her standard of perfection and fitness to continue their kind by reproduction.

From the unremitting operation of this law acting in concert with the tendency which the progeny have to take the more particular qualities of the parents, together with the connected sexual system in vegetables, and instinctive limitation to its own kind in animals, a considerable uniformity of figure, colour, and character, is induced, constituting species; the breed gradually acquiring the very best possible adaptation of these to its condition which it is susceptible of, and when alteration of circumstance occurs, thus changing in character to suit these as far as its nature is susceptible of change.

This circumstance-adaptive law, operating upon the slight but continued natural disposition to sport in the progeny (seedling variety), does not preclude the supposed influence which volition or sensation may have over the configuration of the body. To examine into the disposition to sport in the progeny, even when there is only one parent, as in many vegetables, and to investigate how much variation is modified by the mind or nervous sensation of the parents, or of the living thing itself during its progress to maturity; how far it depends upon external circumstance, and how far on the will, irritability and muscular exertion, is open to examination and experiment. In the first place, we ought to investigate its dependency upon the preceding links of the particular chain of life, variety being often merely types or approximations of former parentage; thence the variation of the family, as well as of the individual, must be embraced by our experiments.

This continuation of family type, not broken by casual particular aberration, is mental as well as corporeal, and is exemplified in many of the dispositions or instincts of particular races of men. These innate or continuous ideas or habits, seem proportionally greater in the insect tribes, those especially of shorter revolution; and forming an abiding memory, may resolve much of the enigma of instinct, and the foreknowledge which these tribes have of what is necessary to completing their round of life, reducing this to knowledge, or impressions, and habits, acquired by a long experience. This greater continuity of existence, or rather continuity of perceptions and mpressions, in insects, is highly probable; it is even difficult in some to ascertain the particular stops when each individuality commences, under the different phases of egg, larva, pupa, or if much consciousness of individuality exists. The continuation of reproduction for several generations by the females alone in some of these tribes, tends to the probability of the greater continuity of existence, and the subdivisions of life by cuttings, at any rate must stagger the advocate of individuality.

Among the millions of *specific varieties* of living things which occupy the humid portion of the surface of our planet, as far back as can be traced, there does not appear, with the exception of man, to have been any particular engrossing race, but a pretty fair balance of powers of occupancy,—or rather, most wonderful variation of circumstance parallel to the nature of every species, as if circumstance and species had grown up together. There are indeed several races which have threatened ascendency in some particular regions, but it is man alone from whom any general imminent danger to the existence of his brethren is to be dreaded.

As far back as history reaches, man had already had considerable influence, and had made encroachments upon his fellow denizens, probably occasioning the destruction of many species, and the production and continuation of a number of varieties or even species, which he found more suited to supply his wants, but which, from the infirmity of their condition—not having undergone selection by the law of nature, of which we have spoken, cannot maintain their ground without his culture and protection.

It is, however, only in the present age that man has

begun to reap the fruits of his tedious education, and has proven how much "knowledge is power." He has now acquired a dominion over the material world, and a consequent power of increase, so as to render it probable that the whole surface of the earth may soon be overrun by this engrossing anomaly, to the annihilation of every wonderful and beautiful variety of animated existence, which does not administer to his wants principally as laboratories of preparation to befit cruder elemental matter for assimilation by his organs.

[*Editor's Note:* Material has been omitted at this point.]

11

Reprinted from *Mag. Nat. Hist. Jour. Zool., Bot., Mineral., Geol., Meteorol.*
8:40–53 (1835)

ART. IV. *An Attempt to classify the "Varieties" of Animals, with Observations on the marked Seasonal and other Changes which naturally take place in various British Species, and which do not constitute Varieties.* By Mr. EDWARD BLYTH.

THE appellation "variety" being very commonly misapplied to individuals of a species, which are merely undergoing a regular natural change, either progressing from youth to maturity, or gradually shifting, according to fixed laws, their colours with the seasons, I conceive that it will be useful to some, to point out a few of the less generally known changes which naturally take place in various British animals; some few of which appear to have been hitherto overlooked, and others to have been described incorrectly.

The term "variety" is understood to signify a departure from the acknowledged type of a species, either in structure, in size, or in colour; but is vague in the degree of being alike used to denote the slightest individual variation, and the most dissimilar breeds which have originated from one common stock. The term is, however, quite inapplicable to an animal in any state of periodical change natural to the species to which it belongs.

Varieties require some classification; and though I feel myself hardly adequate to the task, I shall here propose to arrange them under four principal heads; in the hope that this endeavour will induce some naturalists, more competent than myself, to follow out this intricate and complicated subject, into all its details.

I would distinguish, then, what are called varieties, into *simple variations*, *acquired variations*, *breeds*, and *true varieties*. These appear, in general, sufficiently distinct, although the exact limits of each are sometimes very difficult to be assigned. Indeed, in many cases they only differ in degree, and in others they may be all combined in one individual. Moreover, the varieties of either class have a much greater tendency to produce varieties of another class, than the typical animals of a species have to produce any sort of variety.

I. *Simple Variations.* — The first class, which I propose to style *simple or slight individual variations*, differs only in degree from the last, or *true varieties;* and consists of mere differences of colour or of stature, unaccompanied by any remarkable structural deviation; also of slight individual peculiarities of any kind, which are more or less observable in all animals, whether wild or tame, and which, having a tendency to perpetuate themselves by generation, may, under particular circumstances, become the origin of true *breeds* (which constitute my third class of varieties), but which, in a state of nature, are generally lost in the course of two or three generations. Albinoes belong to this first division, and also the other numerous anomalies mentioned in VII. 589—591. 593—598. These *simple variations* occur both in wild and in domestic animals, but are much more frequent in the latter, and are commonly observed in all *breeds* and *true varieties.*

Among the Mammàlia, total or partial absence of colour is always, I believe, continued through life; excepting, of course, the cases of mere seasonal change; and, in this class of animals generally, perfect albinoes are much more numerous than among birds. Perfect albinoes are peculiar to warm-blooded animals, and in them there is a total deficiency of colouring matter in the rete mucosum, and, consequently, in the fur, and even the pigmentum nigrum of the eye is entirely wanting. In birds, these *perfect* albinoes are extremely rare, although several instances have been recorded in VII. 593. — 598. There are three sorts, however, of true permanent albinoes, which may be thus designated: — 1. *Perfect Albinoes;* which are entirely white, and in which the eyes appear crimson, from the total want of colouring matter, rendering the minute bloodvessels visible: 2. *Semi-Albinoes;* which are either white or of a pale colour all over, and in which the irides are always paler than usual, and not unfrequently blue [I. 66. 178.]: and, 3. *Partial Albinoes;* which are partly of the natural colour, but are more or less mottled *permanently* with white; and in which, if a white patch surrounds the eye, the pigmentum of that organ is commonly wanting. I have thus observed a rabbit, one eye of which was red, and the other dark hazel; but such instances are of very rare occurrence, although (and it is a curious fact) rabbits are often seen wholly white, with the exception of a small patch around each eye; which organ, consequently, is of the usual dark colour. Albinoes, when paired together, as is well known, produce chiefly albino offspring, and a *breed* of them may thus be perpetuated; but, even in a domestic state, they not unfrequently produce young of the usual colour; and, if paired with an ordinary individual, they sometimes produce partial albinoes, or semi-albinoes [I. 178.], and occasionally, if the original colour be brown (as in the case of mice or rabbits*), a black, sandy, or slate-coloured offspring, or an individual with one of these colours more or less varied with white, is produced; but, in the majority of instances, the young wholly resemble one of their parents, and the *preponderance* is decidedly in favour of the natural hue. The coloured offspring of an albino, however, even if matched with another coloured individual, has still a tendency to produce albinoes†, and this fact has been noticed in the human species; but, as Mr. Lawrence observes on the subject (in his *Lectures on the Physiology, Zoology, and the Natural History of Man*), "the disposition to change is 'generally' exhausted in one individual, and the characters of the original stock return, unless the variety is kept up by the precaution above mentioned, of excluding from the breed all which have not the new characters. Thus, when African albinoes intermix with the common race, the offspring generally is black," &c. These observations apply alike to all *simple* or individual *variations*, and to most other varieties, and afford one of many reasons why marked

* These observations are chiefly deduced from the results of some experiments with mice and rabbits.

† Of seven young rabbits thus produced, two were albinoes, one black, and the remainder of the usual colour.

breeds are in a state of nature so rarely perpetuated. There is yet, however, before quitting this subject, another sort of albino to be considered, which, I believe, is peculiar to the feathered race, and which is not, like the others, permanent: these, therefore, I shall denominate *temporary* albinoes. Most of the pale, white, and pied varieties of birds, which are produced in a state of nature, are of this kind. A friend informs me that a perfectly white lark in his possession moulted, and became of the ordinary hue. I lately shot a sparrow which was all over of a very pale brown, or cream colour; it was moulting, and some of the new feathers that were coming were of the usual colour, and others were of a pure white: on the next moult, probably, no more white feathers would have appeared. Of a brood of young robins which frequented my garden, two were white, one partially so, and one of the usual mottled brown; these all moulted into the ordinary colour. I could add other instances to the list, especially amongst domestic poultry. But it does not hence follow that among wild birds there are *no* permanently white or pied varieties; or, in other words, no true partial and semi-albinoes. A blackbird with a white head has now inhabited a garden in this neighbourhood for three successive years; and if the cupidity of collectors did not mark out every white or pied bird for destruction, I doubt not that I should have been able to have furnished some other similar instances of *permanent* variation.

II. *Acquired Variations.* — The second class of varieties which I would designate thus, comprises the various changes which, in a single individual, or in the course of generations, are *gradually* brought about by the operation of known causes: such as the greater or less supply of *nutriment;* the influence of particular *sorts* of food; or, either of these combined with the various privations consequent upon *confinement;* which changes would as gradually and certainly disappear if these causes were removed.

Redundance or deficiency of nutriment affects chiefly the stature of animals. Those herbivorous quadrupeds which browse the scanty vegetation on mountains are invariably much smaller than their brethren which crop the luxuriant produce of the plains; and although the cattle usually kept in these different situations are of diverse breeds; yet either of the breeds gradually removed to the other's pasture, would, in two or three generations, acquire many of the characters of the other, would increase or degenerate in size, according to the supply of nutritious food; though, in either case, they would most probably soon give birth to *true varieties* adapted to the change. In this instance, *temperature* appears only to exert a secondary influence. The Iceland breed of sheep, which feeds on the nutritious lichens of that island, is of large size; and, like the other ruminant animals which subsist on similar food, is remarkable for an extraordinary developement of horns. Another example of *acquired variation*, dependent solely on the supply of nutriment, may be observed in the deciduous horns of the deer family, which are well known to be large or small according to the quality of their food. That *temperature* also does exert an influence greater or less, according to the species of animal, is very evidently shown in the case of the donkey *, of which there are no breeds, nor true varieties, and but very few simple variations [VII. 590.]: this animal is every where found large or small, according to the *climate* it inhabits.

The influence of particular *sorts* of food may be exemplified by the well-known property of madder (*Rŭbia* tinctŏrum), which colours the secretions, and tinges even the bones of the animals which feed on it of a blood-red colour; and, as another familiar instance, may be cited the fact, equally well known, of bullfinches, and one or two other small birds, becoming wholly black when fed entirely on hempseed. I have known, however, this change to take place in a bird (the small aberdevatt finch, so common in the shops), which had been wholly fed on canary seed; yet this by no means invalidates the fact, so often observed, of its being very frequently brought about by the direct influence of the former diet. In several instances which have fallen under my own observation, feeding only on hempseed has invariably superinduced the change.†

The most remarkable of acquired variations are those brought about in animals in a state of confinement or domestication: in which case an animal is supplied regularly with abundance of very nutritious, though often unnatural, food, without the trouble and exertion of having to seek for it, and it becomes, in consequence, bulky and lazy, and in a few generations often very large; while the muscles of the organs of locomotion, from being but little called into action, become rigid and comparatively powerless, or are not developed to their full size. The common domestic breeds of the rabbit, ferret, guinea-pig, turkey, goose, and duck, are thus probably only acquired variations, which, from the causes

* For some curious remarks on this subject, see the excellent article "Ass" in Partington's *Cyclopædia of Natural History*.

† I have not heard, however, that wild bullfinches, hawfinches, and other birds liable to be thus affected, are more commonly found black in localities where hemp is much grown. Amongst others, the skylark and woodlark are very susceptible of the influence of this food.

above-mentioned, have, in the course of generations, become much larger and heavier (excepting, however, in the case of the turkey) than their wild prototypes, and less fitted for locomotion; but which, if turned loose into their natural haunts, would most probably return, in a very few generations, to the form, size, and degree of locomotive ability proper to the species when naturally conditioned.* The crested varieties of domestic geese and ducks, and the hook-billed variety of the latter, are, however, in all probability, *true varieties;* and what are called "lob-eared" rabbits may be either a *true variety*, or a *breed.* The various slight diversities, which I call *simple variations*, are very common in the present class of varieties; and there is also in them a great tendency to produce what I call *true varieties*, as well as these slighter deviations, which, by particular management, may be increased into the sort of variety I denominate *breeds.*

III. *Breeds* are my third class of varieties; and though these may possibly be sometimes formed by accidental isolation in a state of nature, yet they are, for the most part, artificially brought about by the *direct* agency of *man.* It is a general law of nature for all creatures to propagate the like of themselves: and this extends even to the most trivial minutiæ, to the slightest individual peculiarities; and thus, among ourselves, we see a family likeness transmitted from generation to generation. When two animals are matched together, each remarkable for a certain given peculiarity, no matter how trivial, there is also a decided tendency in nature for that peculiarity to *increase;* and if the produce of these animals be set apart, and only those in which the same peculiarity is most apparent, be selected to breed from, the next

* [*A Tame Duck which flies with the same Power, and at the same Height, as a Crow.* (H. S., in I. 378.) — Was not this duck a wild one? I am led to ask this question from having myself witnessed a similar instance. I had often seen a duck, which I had taken to be a tame one, flying about, and always returning to the farm to which it belonged. On enquiry, I found that this duck had been taken, when a duckling, from the nest of a wild duck, and began to fly as soon as it was full grown. The case which H. S. mentions might probably be accounted for in the same manner, as it is by no means likely that so unwieldy a bird as the tame duck should think of trying its wings, after its ancestors had for so many successive generations been satisfied with walking and swimming, and fly "with the same power, and at the same height, as a crow." — *W. H. H. Postmark, Burton on Trent, Oct.* 8. 1834.

The late Rev. Lansdown Guilding had remarked as follows on the case stated by H. S.: —

Domestic birds, from flying little, have their muscles relaxed, or, perhaps, they never acquire their natural strength, for want of exercise. I have observed the geese in Worcestershire, in harvest time, to take very long flights; but, though they went on boldly, they never ascended very far into the air. — *Lansdown Guilding. St. Vincent, May* 1. 1830.]

generation will possess it in a still *more* remarkable degree; and so on, till at length the variety I designate a *breed*, is formed, which may be very unlike the original type.

The examples of this class of varieties must be too obvious to need specification: many of the varieties of cattle, and, in all probability, the greater number of those of domestic pigeons, have been generally brought about in this manner. It is worthy of remark, however, that the original and typical form of an animal is in great measure kept up by the same identical means by which a true *breed* is produced. The original form of a species is *unquestionably* better adapted to its *natural* habits than any modification of that form; and, as the sexual passions excite to rivalry and conflict, and the stronger must always prevail over the weaker, the latter, in a state of nature, is allowed but few opportunities of continuing its race. In a large herd of cattle, the strongest bull drives from him all the younger and weaker individuals of his own sex, and remains sole master of the herd; so that all the young which are produced must have had their origin from one which possessed the maximum of power and physical strength; and which, consequently, in the struggle for existence, was the best able to maintain his ground, and defend himself from every enemy. In like manner, among animals which procure their food by means of their agility, strength, or delicacy of sense, the one best organised must always obtain the greatest quantity; and must, therefore, become physically the strongest, and be thus enabled, by routing its opponents, to transmit its superior qualities to a greater number of offspring. The same law, therefore, which was intended by Providence to keep up the typical qualities of a species, can be easily converted by man into a means of raising different varieties; but it is also clear that, if man did not keep up these breeds by regulating the sexual intercourse, they would all naturally soon revert to the original type. Farther, it is only on this principle that we can satisfactorily account for the degenerating effects said to be produced by the much-censured practice of "breeding in and in."* There would almost seem, in some species, to be a tendency, in every separate family, to some particular kind of deviation; which is only counteracted by the various crossings which, in a state of nature, must take place, and by the above-mentioned law, which causes each race to be chiefly propagated by the most typical and perfect individuals.

V. *True Varieties.* — The last of these divisions to which

* See, however, a good practical article on this subject, entitled "Breeding," in one of the forthcoming numbers of the now publishing edition of Miller's *Dictionary of Gardening and Rural Economy.*

I more peculiarly restrict the term *variety*, consists of what are, in fact, a kind of deformities, or monstrous births, the peculiarities of which, from reasons already mentioned, would very rarely, if ever, be perpetuated in a state of nature; but which, by man's agency, often become the origin of a new race. Such, for example, is the breed of sheep, now common in North America, and known by the name of *ancons* *, or *otter* sheep. A ewe produced a male lamb of peculiar form, with a long body, and short and crooked limbs: the offspring of this animal, with ordinary females, was found sometimes to resemble the one parent, and sometimes the other; but did not usually blend the characters of each; and, in the cases of twins, the two lambs were often equally diverse with their parents. This variety was extensively propagated, in consequence of being less able to jump over fences than the ordinary breeds of sheep. The solidungular [" donkey-footed "] variety of swine, tailless cats, back-feathered, five-toed, and rumpless fowls, together with many sorts of dogs, and probably, also, the race of fan-tailed pigeons, are other striking examples of *true varieties*.

The deviations of this kind do not appear to have any tendency to revert to the original form: this, most probably, could only be restored, in a direct manner, by the way in which the variety was first produced.

To this class may be also referred, with more than probability, some of the more remarkable varieties of the human species. With regard to colour, we know that temperature exerts no *permanent* gradual influence whatever: white races remain unchanged at slight elevations within the tropics; and the natives of Boothia Felix are very dark; the swarthy inhabitants of Mauritania are a white race, and their sunburnt hue is merely an *acquired variation*, which is not transmissible by generation, and which does not extend to those parts which are not exposed to the sun. The colouring principle of black races is inherent in them, and is quite independent of external agency; is even darkest in some parts which are the least exposed, and *vice versâ*. The Ethiopian race is nowhere more black than in the vicinity of the Cape of Good Hope, where the crops are sometimes injured by the winter's frost. Strangely enough, this invariableness of colour constitutes about, perhaps, the most fixed character of these races.

There is one fact, however, here to be observed, which is very well worthy of attention; and this is, that coloured

* *Agkōn*, an elbow, from the crooked form of the forelegs. See Lawrence's *Lectures*, p. 447, 448.

varieties appear to have been chiefly produced in hot countries; which seems almost to induce the conclusion that they were originally efforts of nature, to enable the skin to withstand the scorching produced by exposure to the burning rays of a tropical sun.* How far the structural peculiarities of the negro and other races may not, in some cases, be the effects of *breed*, it would be impossible, perhaps, now to ascertain, and would be worse than presumption, in a novice like myself, to try to determine. Wherever a black individual was produced, especially among rude nations, if the breed was continued at all, the natural aversion it would certainly inspire would soon cause it to become isolated, and, before long, would, most probably, compel the race to seek for refuge in emigration. That no example, however, of the first production of a black variety has been recorded, may be ascribed to various causes; it may have only taken place once since the creation of the human race, and that once in a horde of tropical barbarians remote from the then centres of comparative civilisation, where no sort of record would have been preserved. But it is highly probable that analogous-born varieties may have given rise to the Mongolian, Malay, and certain others of the more diverse races of mankind; nay, we may even suppose that, in some cases, the difference, in the first instance, was much *greater*, and was considerably modified by the intermixture which must have taken place in the first generations. The mixed offspring of two different varieties of man thus generally blends the characters of each; though instances are not wanting of its *entirely* resembling (like the mixed produce of an ancon sheep) either one or the other of its parents; but in this case (as in the albino) the perfect characters of the other parent frequently show themselves in the next generation. I am entering, however, into a wide field, already well trodden by many philosophers; and the subject is already probably pretty well understood by the great majority of readers. Those who are not so familiar with it, will find it ably treated in various works; especially in Dr. Pritchard's work on man, and in the published *Lectures on the Natural History of Man*, by Lawrence: some sound and excellent remarks on *varieties* will also be found in the second volume of Lyell's *Principles of Geology*.

Still, however, it may not be impertinent to remark here, that, as in the brute creation, by a wise provision, the typical characters of a species are, in a state of nature, preserved by

* See Dr. Stark "on the influence of colour on heat and odours," in Jameson's *Edinburgh Journal* for July, 1834; also Professor Powell's reply to it, in the number for October, 184.

those individuals chiefly propagating, whose organisation is the most perfect, and which, consequently, by their superior energy and physical powers, are enabled to vanquish and drive away the weak and sickly, so in the human race degeneration is, in great measure, prevented by the innate and natural preference which is always given to the most comely; and this is the principal and main reason why the varieties which are produced in savage tribes, must generally either become extinct in the first generation, or, if propagated, would most likely be left to themselve, and so become the origin of a new race; and in this we see an adequate cause for the obscurity in which the origin of different races is involved. In a civilised state of society there are other inducements, besides personal attractions, and a new variety in this case, unless very *outré* indeed, would be gradually merged, and in a few generations would disappear entirely by intermixture with the common race. The inferior animals appear not to have the slightest predilection for superior personal appearance; the most dissimilar varieties of the same species mix as freely and readily together as the most typical individuals; the most powerful alone becomes the favourite. Instances of this are not rare in the breeds of dogs.

The above is confessedly a hasty and imperfect sketch, a mere approximation towards an apt classification of "varieties;" but if it chance to meet the eye, and be fortunate enough to engage the attention, of any experienced naturalist, who shall think it worth his while to follow up the subject, and produce a better arrangement of these diversities, my object in indicting the present article will be amply recompensed.

Here, however, I may observe, that the classification I have proposed for specific deviations in the animal creation, is equally applicable to those of the vegetable. The "varieties" in both are strictly analogous.

I come now to the second division of my subject, which is to point out

Some Periodical and other Changes of Appearance, which naturally take place in various British Animals, and which do not constitute Varieties. — Among our native Mammàlia, I know of three principal modes whereby a change of colour is brought about; namely, an actual shedding of the coat; a partial shedding of the coat; and an actual change of colour in the coat itself.

I. As an example of change of appearance produced by actual shedding of the coat, may be instanced the fallow deer

(*Cèrvus Dàma*), whose white spots disappear with the annual casting of its coat in autumn.

II. Partial shedding of the coat takes place in those animals which acquire in autumn a covering of two different kinds: one long, downy, and warm, which is shed in spring; the other short and glossy, which is retained. This change of appearance is exemplified in the common water shrew (*Sòrex* fòdiens), the short summer coat of which is much blacker than the longer downy covering which conceals this in winter. In this little animal the additional winter coat is shed about the latter end of March, or beginning of April; and does not take place uniformly, but progressively, beginning on the head, and ceasing at the hinder extremities; and exhibiting in its progress, throughout, a well-defined line of separation. Animals which (as the British *M*ustelìnæ) have two sorts of fur, the *shorter* of which is the more warm and downy, do not undergo this change, but retain both sorts throughout the year. In these the young have only one kind, which is close and woolly; as is well exemplified in the common polecat (Putòrius Fùro), the young of which are of a very uniform dark brown, and very unlike the old animals.

III. Actual change of colour in the coat itself is exhibited in the appearance of the fallow deer's white spots in spring, and in the case of the mountain hare (*L*èpus variábilis), which is in summer grey, adapted to the hue of the lichens on which it squats; and in winter white, hardly to be discerned upon the snow. The same change also takes place in the stoat or ermine (Putòrius ermíneus), although this is doubted by Mr. Berry (VII. 591.).* In mild winters, such as we have of late experienced in the South of England, but few of the stoats become white, and some of these not until the latter part of the season. The change takes place quickly, but not uniformly, the animal assuming for a short time a pied appearance; but I have not succeeded in ascertaining whether it is accelerated by sudden cold, as the animals are not always to be seen exactly when we want them. One perfectly changed, however, was seen in this neighbourhood soon after the one or two days of very cold weather in the beginning of last

* This gentleman should have mentioned, in his account of the white stoats seen in summer, whether the tail was white or black. If the former, they were doubtless albinoes; if the latter, some constitutional debility may have prevented them from resuming their natural hues. I have seen white stoats late in March, but never after this. Both in these and in the white ferret (a domestic albino variation of the polecat) a decided tinge of yellow is always more or less noticeable.

October. In reference to Mr. Berry's communication, I may observe, that in many dozens of stoats which I have seen in summer, I have never yet seen a white one; whereas in winter, I have seen in the same neighbourhoods a considerable number of white stoats. Where the climate is more excessive, and the transitions of the seasons are more sudden, this change is much more likely to take place generally. In the fur countries, the ermine's change of hue is, I believe, most regular.

There has been, strangely enough, a difference of opinion among naturalists, as to whether these seasonal changes of colour were intended by Providence as an adaptation to change of temperature*, or as a means of preserving the various species from the observation of their foes, by adapting their hues to the colour of the surface; against which latter opinion it has been plausibly enough argued, that "nature provides for the preyer as well as for the prey." The fact is, they answer *both* purposes; and they are among those striking instances of *design*, which so clearly and forcibly attest the existence of an omniscient great First Cause. Experiment demonstrates the soundness of the first opinion; and sufficient proof can be adduced to show that the other is also sound. Some arctic species are white, which have no enemy to fear, as the polar bear, the gyrfalcon, the arctic eagle-owl, the snowy owl, and even the stoat; and therefore, in these, the whiteness can only be to preserve the temperature of their bodies [VI. 79.]; but when we perceive that the colour of nocturnal animals, and of those defenceless species whose habits lead them to be much exposed, especially to enemies from above, are *invariably* of the same colour with their respective natural haunts, we can only presume that this is because they should not appear too conspicuous to their enemies. Thus, in the eloquent language of Mr. Mudie†, who, however, advocates the first opinion, "the ptarmigan is lichen rock in summer, hoar frost in autumn, and snow in winter. Grouse are brown heather, black game are peat bank and shingle, and partridges are clods and withered stalks, all the year round." So, also, on the Continent, the common red-legged partridge (Erýthropus vulgàris) is of the colour of the gravelly and sandy soils on which it is found. So, also, are the different larks, the common quail, the various snipes, and all the other ground squatters, of the hue of their peculiar localities. So, also, are the numerous small Grallatòres which haunt the margin

* See Dr. Stark's paper, before cited, in Jameson's *Philosophical Journal* for July, 1834. [See *M. N. H.*, vi. 79.]

† See Mudie's *Feathered Tribes of the British Islands*, i. 50.

of the ocean, adapted to the colour of the sand. So, also, are those sylvan birds, which quit the dense umbrage of healthy growing trees, to seek their food and expose themselves on bare trunks and leafless decaying branches, of the hue of their particular haunts. "So exquisitely are they fitted for their office," says Mr. Mudie *, "that the several woodpeckers vary in tint with the general colours of the trees which they select. If it is an alternation of green moss, yellow lichen, and ruby tinted cups, with here and there a spot of black, then the green woodpecker comes in charge; but if it is the black and white lichens of the alpine forest or the harsh-juiced tree, then we may look for the spotted races upon the bark." The wryneck is the colour of the lichened branch; and the night swallow and the owls resemble their peculiar places of concealment. So, also, the gayer colours of nocturnal moths are always on the hinder wings†, and the anterior, which, when they rest, conceal these, are adapted to the hues of the various places where by day they are found: even the bright upper wings of the tiger moths (A'rctia Càj*a*, and A. víllica) are with difficulty recognised upon a lichened bank or paling.‡ It is curious, indeed, the resemblance which subsists between the colours of nocturnal birds and night Lepidóptera; the buff tip moth (Pygæ'ra bucéphala) thus reminds us of the barn owl (*S*tríx vulgàris); and the goat moth (*C*óssus Lignipérda), and a host of others, are similar in their tints to most of the *S*trígidæ: in both cases they are doubtless intended for the same purpose, that of concealment. It would indeed be easy to extend this list of examples considerably further; but I shall only now mention the common hare, which, when in form, would hardly ever be seen were it not for its brilliant eye; if its eye were closed, which it probably was before its quick sense of hearing had warned it of our approach, it would almost always, perhaps, wholly escape our observation. This ever continued watchfulness must have given rise to the supposition, that the hare always sleeps with its eyes open.

Seeing, therefore, so many most striking adaptations of colour to haunt, in cases where the concealment thus afforded can be the *only* purpose, I think it is not too much to infer, that the changes of colour in many arctic animals were intended

* See Mudie's *Feathered Tribes of the British Islands*, i. 190.

† Among day-flying Lepidóptera, the more gaudy colours are usually on the *fore* wings.

‡ Animals of bright and gaudy colours are generally very retiring in their habits: even the common robin mostly turns away his breast as you approach.

by Providence for the double purpose of preserving their bodily heat, and of enabling them to elude the observation of their enemies. Certain it is, that their *conspicuousness* would otherwise expose them to inevitable destruction. If I had here space, I could satisfactorily prove that the high-flying Falcónidæ can, in most cases, only perceive their prey when it is *moving*; just, as on the sea-shore, *we* can only distinguish sanderlings when they move. Small Mammàlia which frequent open situations are rarely much abroad, except in the twilight; and ground-feeding birds are ever on the watch, and even the smaller kinds (as I have repeatedly observed) can perceive a hovering falcon *long* before it comes within the sphere of human vision; and they instantly flee to shelter, or they crouch, and lying motionless, so exactly resemble a portion of the surface, that even a hawk's eye cannot distinguish them. Why should the falcon race be endowed with such wonderful powers of enduring hunger and fatigue, if, as is said, at the elevations at which they soar, they can clearly distinguish every living object scattered over the wide expanse beneath them? It is only on such animals as are *off their guard* that they descend; or otherwise, food being so abundant, they would soon multiply to the extirpation of their prey; which, of course, would be very speedily followed by that of the preyer.

How beautifully do we thus perceive, as in a thousand other instances, the balance of nature preserved: and even here we see another reason why sickly or degenerate animals (those, I mean, which are less able to maintain the necessary vigilance) must soon disappear; and why the slightest deviation from the natural hue must generally prove fatal to the animal. How different, thus, are even *simple variations* from the seasonal changes of colour which naturally take place! Properly followed up, this subject might lead to some highly interesting and important results. It certainly points to the conclusion, that every, even the slightest, tint and marking has some decided use, and is intimately connected with the habits and welfare of the animal; and it also furnishes a satisfactory reason, why closely allied animals (or, in other words, animals of very similar form and habits) should so very commonly nearly resemble each other in their colours and in the general character of their markings.

Part V

THE DEVELOPMENT OF THE DARWIN-WALLACE THEORY OF EVOLUTION BY SELECTION: 1831-1858

Editor's Comments on Papers 12 Through 16

The theory that adaptive evolution occurs as the result of selection operating on heritable individual variations was constructed twice, once by Charles Darwin in 1838 (de Beer, 1960–1967), and once by Alfred Russel Wallace in 1858 (Paper 16). Darwin had served as naturalist aboard the H.M.S. *Beagle* during its 1831–1836 world-encircling voyage. He began reading the first volume of Charles Lyell's *Principles of Geology* (1830) before the ship reached the Cape Verde Islands, the first stop on the voyage (Barlow, 1958, p. 77). Darwin was greatly impressed with Lyell's success in applying the uniformity of causes (actualism) principle—the theory that present causes now in operation are sufficient to explain past changes—to scientifically interpret the geological history of the earth. Darwin concluded that Lyell's naturalistic uniformity of causes principle rather than geological catastrophism provided the best theory for interpreting the geological observations that he was making (Barlow, 1958, p. 77).

As Darwin began to prepare his journal of the *Beagle's* voyage for publication and to study his specimen collections, he began to realize that numerous facts indicated the common descent of species. This realization led him to open the first of a series of *Notebooks on Transmutation of Species* (1837–1839, in de Beer, 1960–1967) in which he recorded "any facts that might bear on the question" (Darwin and Seward, 1903, I, p. 367). Darwin made the following entry in his journal for 1837:

> In July opened first note book on "Transmutation of Species"—Had been greatly struck from about Month of previous March on character of S. American fossils—& species on Galapagos Archipelago. These facts origin (especially latter) of all my views. (in de Beer, 1958, p. 25).

Lyell's comprehensive review of the arguments for and against the transmutation of species that first appeared in the second volume (1832) of his *Principles of Geology* provided young Charles Darwin with numerous theories to test his observations against. Darwin read the first edition of Lyell's second volume while on the voyage of the H.M.S. *Beagle* and the fifth edition in 1837. Lyell championed a number of ideas that helped guide Darwin's thoughts as he searched for the natural causes of the transmutation of species. First, Lyell contended that changes that occurred in the past can be explained by reference to causes now in operation (the uniformity of causes principle). Darwin employed uniformitarian methodology in both his geological and biological investigations. Darwin, in an August 29, 1844 letter to L. Horner, wrote:

> the whole merit of the *Principles* was that it altered the whole tone of one's mind and, therefore, that when seeing a thing never seen by Lyell, one yet saw it partially through his eyes . . . [and confessed that] I always feel as if my books came half out of Sir Charles Lyell's brain. . . . (Darwin and Seward, 1903, II, p. 117)

Second, while Lyell opposed the theory that evolution had occurred he did point out that:

> The best authenicated examples of the extent to which species can be made to vary, may be looked for in the history of

> domesticated animals and cultivated plants. . . . [and] that, in perfecting the arts of domesticated animals and cultivated plants, mankind . . . have then been engaged for ages in conducting a series of experiments . . . to ascertain what may be the greatest possible deviation from a common type. . . . (Lyell, 1832, II, see p. 26 in Paper 12)

Lyell's views on the value of studying the selective breeding of domesticated species led Darwin to begin collecting:

> facts on a wholesale scale, more especially with respect to domesticated productions, by printed enquiries, by conversation with skillful breeders and gardeners, and by extensive reading . . . I soon perceived that selection was the keystone of man's success in making useful races of animals and plants. But how selection could be applied to organisms living in a state of nature remained for sometime a mystery to me. (in Barlow, 1958, pp. 119-120)

Lyell discussed the operation of selection in nature but only as a conservative force by which the inferior members of a species were eliminated via the "struggle for existence" that occurred between members of the same species (Paper 12, see pp. 55-56).

Finally, Lyell contended that the extinction of each species was an individual event that was brought about by one or both of two natural processes: (1) change in local circumstances (the physical environment), and (2) the "struggle for existence" between species that can result in the local competitive exclusion—that is, extinction—of a species (Paper 12, pp. 130-132). While Lyell had concluded that the struggle for existence played a negative role leading to the local extinction of populations or even of species, he contended that the "struggle for existence" could not produce evolution because the struggle was too harsh and too immediate due in major part to the invasion of species already fitted to the change of circumstances (Paper 12, pp. 172-175). Charles Darwin used these three theoretical elements of Lyell's thinking— uniformity of causes, argument by analogy from the selective breeding of domesticated organisms, and the importance of the struggle for existence—in constructing his theory of evolution by selection.

The entries that Darwin made in his *Notebooks on Transmutation of Species* (de Beer, 1960-1967) between July 1837 and June 1839 have enabled historians to reconstruct much of the intellectual path that Darwin took in constructing his theory of "descent with modification" by natural, sexual, and artificial selection. For example, Darwin contended in his autobiography that during 1837-1838 when he was trying to identify the natural processes that generated evolution, he "worked on true Baconian principles [inductive logic], and without any theory collected facts on a wholesale scale, more especially with respect to domesticated productions . . . (in Barlow, 1958, p. 119). However, historians who have analyzed the contents of Darwin's *Notebooks on Transmutation* have concluded that, contrary to his recollection, Darwin employed a number of theories to guide him in his search for the natural processes generating evolution (Gruber, 1974; Schweber, 1977; Kohn, 1980). There is even an entry in Darwin's "D" notebook stating that he was employing deductive logic to test the theories of evolution he was considering: "The line of argument often pursued throughout my theory is to establish a point as a probability by induction, to apply it as hypotheses to other points, see whether it will solve them" (in de Beer, 1960–1967, III, p. D117).

The theory of evolution that Darwin was ultimately to construct on September 28, 1838 involved explaining the whole (the evolution of species) in terms of the individual parts (individual organisms) and their interactions (Schweber, 1977, p. 278). What led Darwin to attach such importance to the uniqueness of the individual with respect to the operation of selection on a population of organisms? Was the initial motivation for Darwin's change of focus from the species to the individual the result of his reading the sixth edition of *An Essay on the Principle of Population* by Thomas Malthus (1826) during September, 1838 as Herbert (1971) and others have suggested; or the writings of Dugald Steward and Adam Smith on economics during the summer of 1838 as Schweber (1977, p. 278) has concluded; or the animal breeding literature during the six months before September 28, 1838 as Mayr (1977) has advocated?

Mayr has contended that while Darwin's primary interest in reading the animal breeding literature was to gain a better understanding of the origin of variation,

> in the course of his reading Darwin could not help absorbing the important lesson from the breeders—that every individual in the herd was different from every other one, and what extreme care had to be used in selecting the sires and dams from which to breed the next generation. (Mayr, 1977, p. 325)

Darwin recorded a number of ideas on "picking" (Darwin's term for selecting) in his notebooks (Bajema, 1982). One pamphlet on the selective breeding of animals that he read during the first part of 1838 is thought to have played a crucial role in making him aware of the fact that selective processes analogous to artificial selection were operating in the rest of nature (Ruse, 1975). John Sebright's 1809 pamphlet on *The Art of Improving the Breeds of Domestic Animals* contains not only a review of selective breeding in domestic animals, but, more importantly, a brief discussion of selection operating in the rest of nature. (Excerpts from Darwin's personal annotated copy are presented as Paper 13. The complete text has been reprinted in Bajema, 1982). Sebright gave examples of what Darwin was to later classify as natural selection and sexual selection. Darwin had the analogy between artificial selection and selection in the rest of nature thrust right at him when he read the Sebright pamphlet (Ruse, 1975). He wrote in his second *Notebook on the Transmutation of Species* that the Sebright pamphlet contained: "excellent observations of sickly offspring being cut off so that not propagated by nature . . . and that whole art of making varieties may be inferred from the facts stated" (in de Beer, 1960-1967, II, p. C133).

David Brewster's anonymously published book review of the first two volumes of August Comte's *Cours de Philosophie Positive* that appeared in the July 1838 issue of the *Edinburgh Review* (Brewster, 1838) is thought to have greatly influenced in two important ways the final intellectual steps that Darwin took in constructing his theory of adaptive evolution by selection operating in nature on heritable variations (Schweber, 1977, 1979). First, Brewster pointed out Comte's contention that scientists possess a powerful instrument of investigation, the

> experiment, by means of which we observe bodies out of their natural state; by placing them in artificial aspects and conditions contrived for the purpose of exhibiting to us, under the most favourable circumstances, their phenomena and their properties. Brewster, 1838, p. 303)

This passage helped Darwin to refocus his thinking on the analogy between natural and artificial selection, and to better conceptualize the role that natural selection was to play in the construction of his theory of evolution.

Second, Brewster stressed that Comte attached great importance to "numerical verification," which was "an indispensable criterion . . . of every hypothesis" in the most positive (scientific) of all the sciences, astronomy (Brewster, 1838, p. 299). Darwin, in an effort to make the transmutation theory he was trying to construct as scientific as the theories in physics and astronomy, adopted Comte's criterion and began searching for "numerical verification." In his search for quantitative statements Darwin read an anonymous (1835) book review of the Belgian statistician Adolphe Quetelet's book *Sur l'Homme et le Developpement de ses Facultes* [On man and the development of his faculties]. The anonymous review summarized Thomas Malthus's quantitative predictions on the general question of population:

> Mons. Quetelet agrees very closely with the views of Malthus which he reduces to the following formulae:—"Population tends to increase in a geometrical ratio. The sum of the obstacles, which are opposed to this tendency, is caeteris paribus, as the square of the rapidity of actual increase,"—another instance of the analogy often found to subsist between mechanical laws and

> those which govern human action. "Never, therefore, can population advance so rapidly as to strike with violence upon its utmost possible limit. In approaching that limit, the obstacles must multiply too rapidly to admit of a shock. Nature will, it is true, levy her tribute of deaths in proportion to the nearness of approach; but, that debt being paid in detail, it will be less sensible than if levied at once." (Anonymous, 1835, p. 613)

Darwin had discovered a quantitative statement that he could use in his attempts to construct a theory of evolution. He proceeded to borrow his brother's copy of the sixth edition (1826) of Malthus's *Essay on the Principle of Population.* Charles Darwin was finally able to conceive of selection as a force powerful enough to produce adaptive evolution in nature when he read the first chapter (Paper 15A). There he read Malthus's summary of Benjamin Franklin's (1706–1790) views on "the constant tendency of all animated life to increase beyond nourishment," and on the amount of time it took for the North American British colonial population to double in numbers in the absence of immigration, originally published in Franklin's 1755 essay on *Observations Concerning the Increase of Mankind, Peopling of Countries, etc.* (Paper 14). In response to reading Malthus's summary of Franklin's quantitative statement that ". . . the [human] population, when unchecked, goes on doubling itself every twenty-five years, or increases in a geometric ratio" (Paper 15A, p. 6), Darwin wrote the following entry in his notebook D:

> We ought to be far from wondering of changes in numbers of species, from small changes in nature of locality. Even the energetic language of Decandolle does not convey the warring of the species as inference from Malthus—increase of brutes must be prevented solely by positive checks, excepting that famine may stop desire.—in nature production does not increase, whilst no check prevail, but the positive check of famine & consequently death. I do not doubt every one till he thinks deeply as assumed that increase of animals exactly proportionate to the number that can live.—Population is increase at geometric ratio in FAR SHORTER time than 25 years—yet until the one sentence of Malthus no one clearly perceived the great check amongst men—there is spring like food used for other purposes as wheat for making brandy.—Even a few years plenty, makes population in man increase & an ordinary crop causes a dearth. take Europe on an average every species must have same number killed year with year by hawks, by cold &c.—even one species of hawk decreasing in number must affect instantaneously all the rest.—The final cause of all this wedging, must be to sort out proper structure, & adapt it to changes. —to do that for form, which Malthus shows is the final effect (by means however of volition) of this populousness on the energy of man. One may say there is a force like a hundred thousand wedges trying [to] force every kind

> of adapted structure into the gaps in the oeconomy of nature, or rather forming gaps by thrusting out weaker ones. (Darwin, in de Beer, 1960-1967, III, pp. D134-135)

Darwin now had his theory of evolution by selection ("wedging")!

Malthus, by stating that a population tends to increase in a geometric ratio, enabled Darwin to perceive just how strong a force selection operating in nature could be. The power of selection was generated by the overproduction of offspring that greatly intensified the "struggle for existence" in nature. The contribution that Malthus made to the development of Darwin's ideas concerning the importance of natural selection as an ecological force bringing about "descent with modification" has been succintly summarized by Limoges (1970, p. 79):

> What Malthus provided Darwin was not the idea of the struggle for existence, which was by then common. But rather the notion of the intensity of that struggle, of its constraining power on living beings, the idea that a geometric increase implies that a constant pressure is exerted on living beings which necessarily engenders among them an incessant war—the ancestral form of what is called "population pressure" in present day genetics. (English translation in Schweber, 1978, p. 324)

While Malthus emphasized the environmental constraints on a population always tending to geometrically expand (Papers 15B and 15C), he did not turn the intensity of the struggle into the driving force of evolutionary adaptation because he explicitly rejected the idea that selection operated on virtually unlimited intraspecific variation (Malthus, 1798, pp. 58, 163-171). In contrast to Malthus, who played down the role of intraspecific struggle in his analysis of population regulation in civilized societies, Darwin recognized the importance of intraspecific struggle—competition between individuals within a species—as a driving force of evolutionary adaptation in all species, including humans living in civilized societies (Bowler, 1976, p. 647).

Now that Darwin had constructed his theory of evolution by selection ("wedging") operating on individual heritable variations, what did he do with it? Darwin wrote in his *Autobiography* that he was so anxious to avoid prejudice with respect to his new theory that he determined for some time not to write even the briefest sketch of it (Barlow, 1958, p. 120). Few of the entries Darwin made in his notebooks after reading Malthus explicitly discuss selection, a term that Darwin never used in his 1837-1839 notebooks. He used words such as "destruction," "preservation," "sift," and "picked" to describe natural selection (in de Beer, 1960-1967, IV, pp. E111, E118).

The first record of Darwin's positive views on the analogy between the way that varieties of domesticated species are made by man and the way that species are produced in nature is contained in the following entry Darwin made in his fourth *Notebook on the Transmutation of Species "E"* during December 1838: "It is a beautiful part of my theory, that domesticated races of organics are made by precisely same means as species—but latter far more perfectly and infinitely slower" (in de Beer, 1960–1967, IV, p. E71).

Darwin (1859, p. 61) wrote in the *Origin* that he adopted the metaphor "Natural Selection" to describe the natural means "by which each slight variation, if useful, is preserved . . . in order to mark its relation to man's power of selection." He may have realized the value of selection as a metaphor in communicating his ideas when he read Alexander Walker's brief summary of selection applied to both domestic animals and human beings (Walker, 1839, pp. 263–265, 314–316) or one of William Youatt's books on domesticated animals, each of which contained passages on selective breeding where Youatt used the term *selection* (Bajema, 1982; Vorzimmer, 1977).

The historian Michael Ruse (1979, p. 179) contends that it was probably Darwin's development of the domestic/natural analogy in the context of scientific justification for natural selection that led Darwin back to the context of discovery with respect to sexual selection some time between 1839 and 1842. Ruse points out that man has selectively bred his domesticated species for two qualities: (1) qualities that aid human livelihood—stronger horses, heavier cows, shaggier sheep, bigger vegetables; and (2) qualities that give humans pleasure, such as combative strength—a fiercer fighting cock, and beauty—a fancier pigeon. Ruse contends that Darwin's natural selection theory is analogous to humans selecting traits that help them survive while Darwin's sexual selection theory is analogous to humans selecting traits that give them pleasure. Darwin tied sexual selection for both "combative strength" and "beauty" very tightly to the analogous forms of artificial selection in the *Origin* (Bajema, 1982). This fact makes it seem highly probable that the domestic/natural analogy played a crucial role both in Darwin's isolating sexual selection as a separate force in nature and in his decision that the natural/sexual selection dichotomy was so important that he should draw explicit attention to it in his theory by creating the metaphor "Sexual Selection" (Ruse, 1979, p. 179).

Darwin waited almost four years before he wrote the first summary of his theory, his *Pencil Sketch of 1842* (in de Beer, 1958). To establish the possibility of his theory, Darwin argued by analogy from man's selection that produces varieties under domestication to the circumstances that lead to the production of wild varieties in nature

(Manier, 1978, p. 7). He drew an analogy between artificial selection and both of the forms of selection he perceived to be operating in the rest of nature—natural selection, which he described as "selection by death," and sexual selection, which he described as "selection in time of fullest vigour" among males that involved a "struggle of war or charms" for mates" (Darwin, in de Beer, 1958, pp. 48–49). Two years later Darwin expanded his *Sketch* into his *Essay of 1844* (in de Beer, 1958, pp. 89–254) summarizing what he described in a letter as "the way in which new varieties become so exquisitely adapted to the external conditions of life and to other surrounding beings" (Darwin, 1899, I. p. 395). Darwin made arrangements for the publication of his *Essay of 1844* in case of his death.

In 1854 Darwin finally began preparing for publication his notes on his theory concerning the origin of species by selection. His receipt of Alfred Russel Wallace's manuscript "On the Tendency of Varieties to depart indefinitely from the Type" in June 1858 triggered a chain of events that led to the presentation of three papers at the July 1, 1858, meeting of the Linnean Society (Moody, 1971) and their subsequent publication in the September 1858 *Proceedings of the Linnaean Society, Zoology* (Paper 16). Pertaining to "the Laws which affect the Production of Varieties, Races, and Species" (p. 45), Paper 16 is comprised of an extract from the *Essay of 1844* by Darwin, an abstract from a letter by Darwin to Professor Asa Gray in 1857, and the 1858 essay by Wallace that he had sent to Darwin.

In a June 18, 1858, letter to Charles Lyell, Darwin remarked on the similarity of Wallace's essay to his own version of the theory: "if Wallace had my MS. sketch written out in 1842, he could not have made a better short abstract! Even his terms now stand as heads of my chapters" (Darwin, 1899, I, p. 473). He also noted in a June 25, 1858, letter to Charles Lyell that he differed from Wallace in: ". . . that I was led to . . . my views from what artificial selection has done for domesticated animals" (Darwin, 1899, I, p. 474). Darwin also differed from Wallace in that Darwin coined two metaphors—natural selection and sexual selection— to describe the processes operating in nature that both Wallace and he contended were responsible for varieties departing indefinitely from the original type.

The English naturalist Alfred Russel Wallace (1823–1913) also constructed the theory that adaptive evolution occurs as the result of selection. In 1855 Wallace, who had begun working systematically on the species problem in the 1840s, advocated an evolutionary intrepretation of the geographical distribution of living and extinct species. He wrote "Every species has come into existence coincident in both space and time with a pre-existing closely allied species" (Wallace, 1855, p. 186; see also Pantin, 1959; McKinney, 1972). During one of his bouts with malaria in February 1858, Wallace developed the theory independently of Charles Darwin. Wallace recalled forty years later that:

> During one of these fits, while again considering the problem of the origin of species, something led me to think of Malthus' *Essay on Population* (which I had read about ten years before), and the "positive checks"—war, disease, famine, accidents, etc.—which he adduced as keeping all savage populations nearly stationary. It then occurred to me that these checks must also act upon animals, and keep down their numbers; and as they increase so much faster than man does, while their numbers are always very nearly or quite stationary, it was clear that these checks in their case must be far more powerful, since a number equal to the whole increase must be cut off by them every year. While vaguely thinking how this would affect any species, there suddenly flashed upon me the idea of *survival of the fittest*—that the individuals removed by these checks must be on the whole, *inferior* to those that survived. Then, considering the *variations* continually occurring in every fresh generation of animals or plants, and the changes of climate, of food, of enemies always in progress, the whole method of specific modification became clear to me, and in the two hours of my fit I had thought out the main points of the theory. That same evening I sketched out the draft of a paper; in the two succeeding evenings I wrote it out, and sent it by the next post to Mr. Darwin. (Wallace, 1898, p. 139)

Wallace (1908) also wrote a note in which he quoted some of Malthus's passages as being fairly illustrative of those which he recalled in February 1858 when thinking about how the populations of human savages are kept down (Papers 15B and 15C). He recalled that "along with Malthus I had read, and been even more impressed by, Sir Charles Lyell's immortal *Principles of Geology*" (Wallace, 1908, p. 118). An excellent biography of Wallace's life and scientific approach to the species problem up to 1858 has been published by McKinney (1972).

Charles Darwin, still very concerned about who would be given priority with respect to the discovery of the theory of adaptive evolution by selection, began abstracting his long species manuscript and published his abstract in 1859 under the title *On the Origin of Species by Means of Natural Selection, or the Preservation of Favoured Races in the Struggle for Life.*

REFERENCES

Anonymous, 1835, [Review of] *On Man, and the Development of his Faculties, &c.* [*Sur l'Homme et le Developpment de ses Facultes, &c.*] by A. Quetelet, *Athenaeum,* part I, pp. 593-595; part II, pp. 611-613; part III, pp. 658-661.

Bajema, C. J., ed., 1982, *Artificial Selection and the Development of Evolutionary Theory,* Benchmark Papers in Systematic and Evolutionary Biology, vol. 4, Hutchinson Ross, Stroudsburg, Pa., 384p.

Barlow, N., ed., 1958, *The Autobiography of Charles Darwin,* Harcourt, Brace, and World, New York, 253p. (Original omissions restored, appendix and notes added.)
Bowler, P. J., 1976, Malthus, Darwin, and the Concept of Struggle, *J. Hist. Ideas* **37:**631-650.
Brewster, D. [anonymous], 1838, Review of Comte's *Cours de Philosophie Positive, Edinburgh Rev.* **67:**271-308.
Darwin, C. R., 1859, *On the Origin of Species by Means of Natural Selection or the Preservation of Favoured Races in the Struggle for Life,* 1st ed., J. Murray, London, 502p. (Facsimile edition by Harvard University Press, 1966.)
Darwin, F., 1899, *The Life and Letters of Charles Darwin,* 2 vols., Appleton, New York, 558p., 562p.
Darwin, F., and A. C. Seward, eds., 1903, *More Letters of Charles Darwin,* 2 vols., Appleton, New York, 494p., 508p.
de Beer, G., ed., 1958, *Evolution by Selection,* Cambridge University Press, Cambridge, 288p.
de Beer, G., ed., 1960-1967, *Darwin's Notebooks on Transmutation of Species.*
Part I. First Notebook (July 1837-February 1838) [Notebook B], *Br. Mus. (Nat. Hist.) Bull. Hist. Ser.* **2**(2):23-73 (1960).
Part II. Second Notebook (February 1838-July 1838) [Notebook C], *Br. Mus. (Nat. Hist.) Bull. Hist. Ser.* **2**(3):75-118 (1960).
Part III. Third Notebook (July 15, 1838-October 2, 1838) [Notebook D], *Br. Mus. (Nat. Hist.) Bull. Hist. Ser.* **2**(4):119-150 (1960).
Part IV. Fourth Notebook (October 1838-July 10, 1839) [Notebook E], *Br. Mus. (Nat. Hist.) Bull. Hist. Ser.* **2**(5):151-183 (1960).
Part V. Addenda and Corrigenda, *Br. Mus. (Nat. Hist.) Bull. Hist. Ser.* **2**(6):185-200 (1961).
Part V. Pages Excised by Darwin, *Br. Mus. (Nat. Hist.) Bull. Hist. Ser.* **3**(5):129-176 (1967).
Gruber, H. E., 1974, *Darwin on Man: A Psychological Study of Scientific Creativity,* Dutton, New York, 495p.
Herbert, S., 1971, Darwin, Malthus and Selection, *J. Hist. Biol.* **4:**209-217.
Kohn, D., 1980, Theories to Work By: Rejected Theories, Reproduction, and Darwin's Path to Natural Selection, *Studies Hist. Biol.* **4:**67-170.
Limoges, C., 1970, *La Selection Naturelle. Etude sur la Premiere Constitution d'un concept, 1837-1839,* Presses Universitaires de France, Paris, 184p.
Lyell, C., 1830, *Principles of Geology, Being an Attempt to Explain the Former Changes of the Earth's Surface, by Reference to Causes Now in Operation,* vol. 1, Murray, London, 511p.
Lyell, C., 1832, *Principles of Geology, Being an Attempt to Explain the Former Changes of the Earth's Surface by Reference to Causes Now in Operation,* vol. 2, Murray, London, 330p.
McKinney, H. L., 1972, *Wallace and Natural Selection,* Yale University Press, New Haven, 193p.
Malthus, T. R. [anonymous], 1798, *An Essay on the Principle of Population, As It Affects the Future Improvement of Society. With Remarks on the Speculations of Mr. Godwin, M. Condorcet and Other Writers,* J. Johnson, London, 396p. (Pages 163-171 are reprinted as Paper 6 in Bajema, 1982, pp. 41-49).
Malthus, T. R., 1826, *An Essay on the Principle of Population; or, A View of Its Past and Present Effects on Human Happiness; With an Inquiry into*

Our Prospects Respecting the Future Removal or Mitigation of the Evils Which It Occasions, 6th ed., 2 vols., Murray, London, 535p., 528p.

Manier, E., 1978, *The Young Darwin and His Cultural Circle: A Study of Influences Which Helped Shape the Language and Logic of the First Drafts of the Theory of Natural Selection,* D. Reidel, Boston, 242p.

Mayr, E., 1977, Darwin and Natural Selection, *Am. Sci.* **65:**321-327.

Moody, J., 1971, The Reading of the Darwin and Wallace Papers: An Historical "Non-Event," *J. Soc. Bibliog. Nat. Hist.* **5:**474-476.

Pantin, C., 1959, Alfred Russel Wallace: His Pre-Darwinian Essay of 1855. *Linnean Soc. London Proc.* **171:**139-140.

Ruse, M., 1975, Charles Darwin and Artificial Selection, *J. Hist. Ideas* **36:**339-350.

Ruse, M., 1979, *The Darwinian Revolution: Science Red in Tooth and Claw,* University of Chicago Press, Chicago, 320p.

Schweber, S. S., 1977, The Origin of the *Origin* Revisited, *J. Hist. Biol.* **10:**229-316.

Schweber, S. S., 1978, The Genesis of Natural Selection—1838: Some Further Insights, *BioSci.* **28:**321-326.

Schweber, S. S., 1979, The Young Darwin, *J. Hist. Biol.* **12:**175-192.

Sebright, J., 1809, *The Art of Improving the Breeds of Domestic Animals. In a Letter Addressed to the Right Hon. Sir Joseph Banks, K. B.,* J. Harding, London, 31p.

Vorzimmer, P. J., 1977, The Darwin Reading Notebooks (1838-1860), *J. Hist. Biol.* **10:**107-153.

Walker, A., 1839, *Intermarriage: or The Mode in Which, and the Causes Why, Beauty, Health and Intellect, Result from Certain Unions, and Deformity, Disease, and Insanity, From Others: . . . and . . . an Account of Corresponding Effects in the Breeding of Animals,* J. & H. Langley, New York, 384p.

Wallace, A. R., 1855, On the Law Which has Regulated the Introduction of New Species, *Annu. Mag. Nat. Hist.* **16:**184-196.

Wallace, A. R., 1898, *The Wonderful Century: Its Successes and Failures,* Sonnenschein, London, 412p.

Wallace, A. R., 1908, Notes on the Passages of Malthus's "Principles of Population" Which Suggested the Idea of Natural Selection to Darwin and Myself, in *The Darwin-Wallace Celebration Held on 1st July, 1908, by the Linnaean Society of London,* Linnaean Society, London, pp. 111-118.

12

Reprinted from pages 25-27, 55-56, 130-132, and 172-175 of *Principles of Geology, Being an Attempt to Explain the Former Changes of the Earth's Surface by Reference to Causes Now in Operation,* vol. 2, J. Murray, London, 1832

PRINCIPLES OF GEOLOGY, BEING AN ATTEMPT TO EXPLAIN THE FORMER CHANGES OF THE EARTH'S SURFACE BY REFERENCE TO CAUSES NOW IN OPERATION

C. Lyell

The difference of stature in some races of dogs in comparison to others, is as one to five in linear dimensions, making a difference of a hundred-fold in volume*. Now there is good reason to believe that species in general are by no means susceptible of existing under a diversity of circumstances, which may give rise to such a disparity in size, and consequently, there will be a multitude of distinct species, of which no two adult individuals can ever depart so widely from a certain standard of dimensions as the mere varieties of certain other species,—the dog for instance. Now we have only to suppose that what is true of size, may also hold in regard to colour and many other attributes, and it will at once follow that the degree of possible discordance between varieties of the same species, may in certain cases exceed the utmost disparity which can even arise between two individuals of many distinct species.

The same remarks may hold true in regard to instincts; for if it be foreseen that one species will have to encounter a great variety of foes, it may be necessary to arm it with great cunning and circumspection, or with courage or other qualities capable of developing themselves on certain occasions; such for example as those migratory instincts which are so remarkably exhibited at particular periods, after they have remained dormant for many generations. The history and habits of one variety of such a species, may often differ more considerably from some other than those of many distinct species which have no such latitude of accommodation to circumstances.

Lamarck has somewhat misstated the idea commonly entertained of a species, for it is not true that naturalists in general assume that the organization of an animal or plant remains absolutely constant, and that it can never vary in any of its parts. All must be aware that circumstances influence the habits, and that the habits may alter the state of the parts and

* Cuvier, Disc. Prelim., p. 128, sixth edition.

organs *. But the difference of opinion relates to the extent to which these modifications of the habits and organs of a particular species may be carried.

Now let us first inquire what positive facts can be adduced in the history of known species, to establish a great and permanent amount of change in the form, structure, or instinct of individuals descending from some common stock. The best authenticated examples of the extent to which species can be made to vary, may be looked for in the history of domesticated animals and cultivated plants. It usually happens that those species, both of the animal and vegetable kingdom, which have the greatest pliability of organization, those which are most capable of accommodating themselves to a great variety of new circumstances, are most serviceable to man. These only can be carried by him into different climates, and can have their properties or instincts variously diversified by differences of nourishment and habits. If the resources of a species be so limited, and its habits and faculties be of such a confined and local character, that it can only flourish in a few particular spots, it can rarely be of great utility.

We may consider, therefore, that in perfecting the arts of domesticating animals and cultivating plants, mankind have first selected those species which have the most flexible frames and constitutions, and have then been engaged for ages in conducting a series of experiments, with much patience and at great cost, to ascertain what may be the greatest possible deviation from a common type which can be elicited in these extreme cases.

The modifications produced in the different races of dogs, exhibit the influence of man in the most striking point of view. These animals have been transported into every climate, and placed in every variety of circumstances; they have been made, as a modern naturalist observes, the servant, the companion, the guardian, and the intimate friend of man, and the power of a superior genius has had a wonderful influence, not only on

* Phil. Zool., tom. i. p. 266.

their forms, but on their manners and intelligence*. Different races have undergone remarkable changes in the quantity and colour of their clothing: the dogs of Guinea are almost naked, while those of the Arctic circle are covered with a warm coat both of hair and wool, which enables them to bear the most intense cold without inconvenience. There are differences also of another kind no less remarkable, as in size, the length of their muzzles, and the convexity of their foreheads.

But if we look for some of those essential changes which would be required to lend even the semblance of a foundation for the theory of Lamarck, respecting the growth of new organs and the gradual obliteration of others, we find nothing of the kind. For in all these varieties of the dog, says Cuvier, the relation of the bones with each other remain essentially the same; the form of the teeth never changes in any perceptible degree, except that in some individuals, one additional false grinder occasionally appears, sometimes on the one side, and sometimes on the other†. The greatest departure from a common type, and it constitutes the maximum of variation as yet known in the animal kingdom, is exemplified in those races of dogs which have a supernumerary toe on the hind foot with the corresponding tarsal bones, a variety analogous to one presented by six-fingered families of the human race‡.

[*Editor's Note:* Material has been omitted at this point.]

If we consider the vegetable kingdom generally, it must be recollected, that even of the seeds which are well ripened, the greater part are either eaten by insects, birds, and other animals, or decay for want of room and opportunity to germinate. Unhealthy plants are the first which are cut off by causes prejudicial to the species, being usually stifled by more vigorous individuals of their own kind. If, therefore, the relative fecundity or

* Dureau de la Malle, Ann. des. Sci. Nat. tom. xxi. p. 63. Sept. 1830.

† Disc. Prel., p. 129, sixth edition.

‡ Ibid.

hardiness of hybrids be in the least degree inferior, they cannot maintain their footing for many generations, even if they were ever produced beyond one generation in a wild state. In the universal struggle for existence, the right of the strongest eventually prevails; and the strength and durability of a race depends mainly on its prolificness, in which hybrids are acknowledged to be deficient.

Centaurea hybrida, a plant which never bears seed, and is supposed to be produced by the frequent intermixture of two well-known species of Centaurea, grows wild upon a hill near Turin. *Ranunculus lacerus*, also steril, has been produced accidentally at Grenoble, and near Paris, by the union of two Ranunculi; but this occurred in gardens *.

Mr. Herbert, in one of his ingenious papers on mule plants, endeavours to account for their non-occurrence in a state of nature, from the circumstance that all the combinations that were likely to occur, have already been made many centuries ago, and have formed the various species of botanists; but in our gardens, he says, whenever species, having a certain degree of affinity to each other, are transported from different countries, and brought for the first time into contact, they give rise to hybrid species †. But we have no data, as yet, to warrant the conclusion, that a single permanent hybrid race has ever been formed, even in gardens, by the intermarriage of two allied species brought from distant habitations. Until some fact of this kind is fairly established, and a new species, capable of perpetuating itself in a state of perfect independence of man, can be pointed out, we think it reasonable to call in question entirely this hypothetical source of new species. That varieties do sometimes spring up from cross breeds, in a natural way, can hardly be doubted, but they probably die out even more rapidly than races propagated by grafts or layers.

* Hon. and Rev. W. Herbert, Hort. Trans., vol. iv., p. 41. † Ibid

[*Editor's Note:* Material has been omitted at this point.]

let us turn to another class of facts, and examine attentively the circumstances which determine the *stations* of particular animals and plants, and perhaps we shall discover, in the vicissitudes to which these stations are exposed, a cause fully adequate to explain the phenomena under consideration.

Stations comprehend all the circumstances, whether relating to the animate or inanimate world, which determine whether a given plant or animal can exist in a given locality, so that if it be shown that stations can become essentially modified by the influence of known causes, it will follow that species, as well as individuals, are mortal.

Every naturalist is familiar with the fact, that although in a particular country, such as Great Britain, there may be more than three thousand species of plants, ten thousand insects, and a great variety in each of the other classes, yet there will not be more than a hundred, perhaps not half that number, inhabiting any given locality. There may be no want of space in the supposed tract; it may be a large mountain, or an extensive moor, or a great river-plain, containing room enough for individuals of every species in our island; yet the spot will be occupied by a few to the exclusion of many, and these few are enabled, throughout long periods, to maintain their ground successfully against every intruder, notwithstanding the facilities which species enjoy, by virtue of their powers of diffusion, of invading adjacent territories.

The principal causes which enable a certain assemblage of plants thus to maintain their ground against all others depend, as is well known, on the relations between the physiological nature of each species, and the climate, exposure, soil, and other physical conditions of the locality. Some plants live only on rocks, others in meadows, a third class in marshes. Of the latter, some delight in a fresh-water morass,—others in salt marshes, where their roots may copiously absorb saline particles. Some prefer an alpine region in a warm latitude, where, during the heat of summer, they are constantly irrigated by the cool waters of melting snows. To others loose sand, so fatal to the generality of species, affords the most proper station. The *Carex arenaria* and the *Elymus arenarius* acquire their full vigour on a sandy dune, obtaining an ascendency over the very plants which in a stiff clay would immediately stifle them.

Where the soil of a district is of so peculiar a nature that it is extremely favourable to certain species, and agrees ill with every other, the former get exclusive possession of the ground, and, as in the case of heaths, live in societies. In like manner, the Bog moss (*Hypnum palustre*) is fully developed in peaty swamps, and becomes, like the heath, in the language of botanists, a social plant. Such monopolies would be very frequent, if the powers of a great number of species were not equally balanced, and if animals did not interfere most actively to preserve an equilibrium in the vegetable kingdom.

"All the plants of a given country," says Decandolle in his usual spirited style, "are at war one with another. The first which establish themselves by chance in a particular spot, tend, by the mere occupancy of space, to exclude other species—the greater choke the smaller, the longest livers replace those which last for a shorter period, the more prolific gradually make themselves masters of the ground, which species multiplying more slowly would otherwise fill."

In this continual strife, it is not always the resources of the plant itself which enable it to maintain or extend its ground. Its success depends, in a great measure, on the number of its foes or allies among the animals and plants inhabiting the same region. Thus, for example, a herb which loves the shade may multiply, if some tree with spreading boughs and dense foliage flourish in the neighbourhood. Another, which, if unassisted, would be overpowered by the rank growth of some hardy competitor, is secure, because its leaves are unpalatable to cattle, which, on the other hand, annually crop down its antagonist, and rarely suffer it to ripen its seed.

[*Editor's Note:* Material has been omitted at this point.]

(172)

Let us now consider more particularly the effect of vicissitudes of climate in causing one species to give way before the increasing numbers of some other.

When temperature forms the barrier which arrests the progress of an animal or plant in a particular direction, the individuals are fewer and less vigorous as they approach the extreme confines of the geographical range of the species. But these stragglers are ready to multiply rapidly on the slightest increase or diminution of heat that may be favourable to them, just as particular insects increase during a hot summer, and certain plants and animals gain ground after a series of congenial seasons.

In almost every district, especially if it be mountainous, there are a variety of species the limits of whose habitations are conterminous, some being unable to proceed farther without encountering too much heat, others too much cold. Individuals, which are thus on the borders of the regions proper to their respective species, are like the out-posts of hostile armies, ready to profit by every slight change of circumstances in their favour, and to advance upon the ground occupied by their neighbours and opponents.

The proximity of distinct climates, produced by the inequalities of the earth's surface, brings species possessing very different constitutions into such immediate contact, that their naturalizations are very speedy whenever opportunities of advancing present themselves. Many insects and plants, for example, are common to low plains within the arctic circle, and to lofty mountains in Scotland and other parts of Europe. If the climate, therefore, of the polar regions were transferred to our own latitudes, the species in question would immediately descend from these elevated stations to overrun the low grounds. Invasions of this kind, attended by the expulsion of the pre-occupants, are almost instantaneous, because the change of temperature not only places the one species in a more favourable position, but renders the others sickly and almost incapable of defence.

Lamarck appears to have speculated on the modifications to which every variation of external circumstances might give rise in the form and organization of species, as if he had indefinite periods of time at his command, not sufficiently reflecting that revolutions in the state of the habitable earth, whether by changes of climate or any other condition, are attended by still greater fluctuations in the relative condition of contemporary species. They can avail themselves of these alterations in their favour instantly, and augment their numbers to the injury of some other species; whereas the supposed transmutations are only assumed to be brought about by slow and insensible degrees, and in a lapse of ages, the duration of which is beyond the reach of human conception Even if we thought it possible that the palm or the elephant, which now flourish in equatorial regions, could ever learn to bear the variable seasons of our temperate zone, or the rigours of an arctic winter, we should, with no less confidence, affirm, that they must perish before they had time to become habituated to such new circumstances.

That they would be supplanted by other species at each variation of climate, may be inferred from what we have before said of the known local exterminations of species which have resulted from the multiplication of others. Some minute insect, perhaps, might be the cause of destruction to the huge and powerful elephant.

Suppose the climate of the highest part of the woody zone of Etna to be transferred to the sea-shore at the base of the mountain, no botanist would anticipate that the olive, lemon-tree, and prickly pear (*Cactus opuntia*), would be able to contend with the oak and chestnut, which would begin forthwith to descend to a lower level, or that these last would be able to stand their ground against the pine, which would also, in the space of a few years, begin to occupy a lower position. We might form some kind of estimate of the time which might be required for the migrations of these plants; whereas we have no data for concluding that any number of thousands of years would be sufficient for one step in the pretended metamorphosis of one species into another, possessing distinct attributes and qualities.

This argument is applicable not merely to *climate*, but to any other cause of mutation. However slowly a lake may be converted into a marsh, or a marsh into a meadow, it is evident that before the lacustrine plants can acquire the power of living in marshes, or the marsh-plants of living in a less humid soil, other species, already existing in the region, and fitted for these several stations, will intrude and keep possession of the ground. So if a tract of salt-water becomes fresh by passing through every intermediate degree of brackishness, still the marine molluscs will never be permitted to be gradually metamorphosed into fluviatile species; because long before any such transformation can take place by slow and insensible degrees, other tribes, which delight in brackish or fresh-water, will avail themselves of the change in the fluid, and will, each in their turn, monopolize the space.

It is idle to dispute about the abstract possibility of the conversion of one species into another, when there are known causes so much more active in their nature, which must always intervene and prevent the actual accomplishment of such conversions. A faint image of the certain doom of a species less fitted to struggle with some new condition in a region which it previously inhabited, and where it has to contend with a more vigorous species, is presented by the extirpation of savage tribes of men by the advancing colony of some civilized nation. In this case the contest is merely between two different *races*, each gifted with equal capacities of improvement—between two varieties, moreover, of a species which exceeds all others in its aptitude to accommodate its habits to the most extraordinary variations of circumstances. Yet few future events are more certain than the speedy extermination of the Indians of North America and the savages of New Holland in the course of a few centuries, when these tribes will be remembered only in poetry and tradition.

13

Reprinted from pages 15-16 of *The Art of Improving the Breeds of Domestic Animals. In a Letter Addressed to the Right Hon. Sir Joseph Banks, K.B.*, J. Harding, London, 1809

THE ART OF IMPROVING THE BREEDS OF DOMESTIC ANIMALS. IN A LETTER ADDRESSED TO THE RIGHT HON. SIR JOSEPH BANKS, K.B.

J. Sebright

[*Editor's Note:* The annotations are those of Charles Darwin. Some were made in pen, others in pencil; hence the variations in density. In the original, material precedes and follows this excerpt.]

15

Many causes combine to prevent animals, in a state of nature, from degenerating; they are perpetually intermixing, and therefore do not feel the bad effects of breeding *in-and-in*: the perfections of some correct the imperfections of others, and they go on without any material alteration, except what arises from the effects of food and climate.

The greatest number of females will, of course, fall to the share of the most vigorous males; and the strongest individuals of both sexes, by driving away the weakest, will enjoy the best food, and the most favourable situations, for themselves and for their offspring.

A severe winter, or a scarcity of food, by destroying the weak and the unhealthy, has all the good effects of the most skilful selection. In cold and barren countries no animals can live to the age of maturity, but those who have strong constitutions; the weak and the unhealthy do

which would be infertile unto the selection —

16

not live to propagate their infirmities, as is too often the case with our domestic animals. To this I attribute the peculiar hardiness of the horses, cattle, and sheep, bred in mountainous countries, more than to their having been inured to the severity of the climate; for our domestic animals do not become more hardy by being exposed, when young, to cold and hunger: animals so treated will not, when arrived at the age of maturity, endure so much hardship as those who have been better kept in their infant state.

14

Reprinted from pages 1–15 appended to *Observations on the late and present Conduct of the French, with Regard to their Encroachments upon the British Colonies in North America. Together with Remarks on the Importance of these Colonies to Great Britain. To which is added, wrote by another Hand; Observations concerning the Increase of Mankind, Peopling of Countries, &c.*, William Clarke, S. Kneeland, Boston, 1755

OBSERVATIONS concerning the Increafe of Mankind, Peopling of Countries, &c.

Benjamin Franklin

1. TAbles of the Proportion of Marriages to Births, of Deaths to Births, of Marriages to the Numbers of Inhabitants, &c. form'd on Obfervaions made upon the Bills of Mortality, Chriftnings, &c. of populous Cities, will not fuit Countries; nor will Tables form'd on Obfervations made on full fettled old Countries, as *Europe*, fuit new Countries, as *America*.

2. For People increafe in Proportion to the Number of Marriages, and that is greater in Proportion to the Eafe and Convenience of fupporting a Family. When Families can be eafily fupported, more Perfons marry, and earlier in Life.

3. In Cities, where all Trades, Occupations and Offices are full, many delay marrying, till they can fee how to bear the Charges of a Family; which Charges are

are greater in Cities, as Luxury is more common: many live ſingle during Life, and continue Servants to Families, Journeymen to Trades,&c. hence Cities do not by natural Generation ſupply themſelves with Inhabitants; the Deaths are more than the Births.

4. In Countries full ſettled, the Caſe muſt be nearly the ſame; all Lands being occupied and improved to the Heighth; thoſe who cannot get Land, muſt Labour for others that have it; when Labourers are plenty, their Wages will be low; by low Wages a Family is ſupported with Difficulty; this Difficulty deters many from Marriage, who therefore long continue Servants and ſingle.—Only as the Cities take Supplies of People from the Country, and thereby make a little more Room in the Country; Marriage is a little more incourag'd there, and the Births exceed the Deaths.

5. *Europe* is generally full ſettled with Huſbandmen, Manufacturers, &c. and therefore cannot now much increaſe in People: *America* is chiefly occupied by Indians, who ſubſiſt moſtly by Hunting.—But as the Hunter, of all Men, requires the greateſt Quantity of Land from whence to draw his Subſiſtence, (the Huſbandman ſubſiſting

ſubſiſting on much leſs, the Gardner on ſtill leſs, and the Manufacturer requiring leaſt of all), The *Europeans* found *America* as fully ſettled as it well could be by Hunters ; yet theſe having large Tracks, were eaſily prevail'd on to part with Portions of Territory to the new Comers, who did not much interfere with the Natives in Hunting, and furniſh'd them with many Things they wanted.

6. Land being thus plenty in *America*, and ſo cheap as that a labouring Man, that underſtands Huſbandry, can in a ſhort Time ſave Money enough to purchaſe a Piece of new Land ſufficient for a Plantation, whereon he may ſubſiſt a Family; ſuch are not afraid to marry; for if they even look far enough forward to conſider how their Children when grown up are to be provided for, they ſee that more Land is to be had at Rates equally eaſy, all Circumſtances conſidered.

7. Hence Marriages in *America* are more general, and more generally early, than in *Europe*. And if it is reckoned there, that there is but one Marriage per Annum among 100 Perſons, perhaps we may here reckon two ; and if in *Europe* they have but 4 Births to a Marriage (many of their Marriages being late) we may here reckon

8,

8, of which if one half grow up, and our Marriages are made, reckoning one with another at 20 Years of Age, our People muſt at leaſt be doubled every 20 Years.

8. But notwithſtanding this Increaſe, ſo vaſt is the Territory of *North-America*, that it will require many Ages to ſettle it fully; and till it is fully ſettled, Labour will never be cheap here, where no Man continues long a Labourer for others, but gets a Plantation of his own, no Man continues long a Journeyman to a Trade, but goes among thoſe new Settlers, and ſets up for himſelf, &c. Hence Labour is no cheaper now, in *Pennſylvania*, than it was 30 Years ago, tho' ſo many Thouſand labouring People have been imported.

9. The Danger therefore of theſe Colonies interfering with their Mother Country in Trades that depend on Labour, Manufactures, &c. is too remote to require the Attention of *Great-Britain*.

10. But in Proportion to the Increaſe of the Colonies, a vaſt Demand is growing for Britiſh Manufactures, a glorious Market wholly in the Power of *Britain*, in which Foreigners cannot interfere, which will increaſe in a ſhort Time even beyond her Power of ſupplying, tho' her whole Trade ſhould be to her Colonies: Therefore *Britain* ſhould

ſhould not too much reſtrainManufactures in her Colonies. A wiſe and good Mother will not do it. To diſtreſs, is to weaken, and weakening the Children, weakens the whole Family.

11. Beſides if theManufactures of *Britain* (by Reaſon of the *American* Demands) ſhould riſe too high in Price, Foreigners who can ſell cheaper will drive her Merchants out of Foreign Markets ; Foreign Manufactures will thereby be encouraged and increaſed, and conſequently foreign Nations, perhaps her Rivals in Power,grow more populous and more powerful ; while her own Colonies, kept too low,are unable to aſſiſt her, or add to her Strength.

12. 'Tis an ill-grounded Opinion that by the Labour of Slaves, *America* may poſſibly vie in Cheapneſs of Manufactures with *Britain.* The Labour of Slaves can never be ſo cheap here as the Labour of working Men is in *Britain.* Any one may compute it. Intereſt of Money is in theColonies from 6 to 10 per Cent. Slaves one with another coſt 30 £. Sterling per Head. Reckon then the Intereſt of the firſtPurchaſe of a Slave, the Inſurance or Riſque on his Life, his Cloathing and Diet, Expences in his Sickneſs and Loſs of Time, Loſs by his Neglect of Buſineſs (Neglect is natural to the

the Man who is not to be benefited by his own Care or Diligence), Expence of a Driver to keep him at Work, and his Pilfering from Time to Time, almoſt every Slave being *by Nature* a Thief, and compare the whole Amount with the Wages of a Manufacturer of Iron or Wool in *England*, you will ſee that Labour is much cheaper there than it ever can be by Negroes here. Why then will *Americans* purchaſe Slaves? Becauſe Slaves may be kept as long as a Man pleaſes, or has Occaſion for their Labour; while hired Men are continually leaving their Maſter (often in the midſt of his Buſineſs,) and ſetting up for themſelves. §. 8.

13. As the Increaſe of People depends on the Encouragement of Marriages, the following Things muſt diminiſh a Nation, *viz.* 1. The being conquered; for the Conquerors will engroſs as many Offices, and exact as much Tribute or Profit on the Labour of the conquered, as will maintain them in their new Eſtabliſhment, and this diminiſhing the Subſiſtence of the Natives diſcourages their Marriages, & ſo gradually diminiſhes them, while the Foreigners increaſe. 2. Loſs of Territory. Thus the *Britons* being driven into *Wales*, and crowded together in a barren Country inſufficient

ſufficient to ſupport ſuch great Numbers, diminiſhed 'till the People bore a Proportion to the Produce, while the *Saxons* increas'd on their abandoned Lands; 'till the Iſland became full of *Engliſh*. And were the *Engliſh* now driven into *Wales* by ſome foreign Nation, there would in a few Years be no more Engliſhmen in *Britain*, than there are now People in *Wales*. 3. Loſs of Trade. Manufactures exported, draw Subſiſtence from Foreign Countries for Numbers; who are thereby enabled to marry and raiſe Families. If the Nation be deprived of any Branch of Trade, and no new Employment is found for the People occupy'd in that Branch, it will alſo be ſoon deprived of ſo many People. 4. Loſs of Food. Suppoſe a Nation has a Fiſhery, which not only employs great Numbers, but makes the Food and Subſiſtence of the People cheaper: If another Nation becomes Maſter of the Seas, and prevents the Fiſhery, the People will diminiſh in Proportion, as the Loſs of Employ, and Dearneſs of Proviſion, makes it more difficult to ſubſiſt a Family. 5. Bad Government and inſecure Property. People not only leave ſuch a Country, and ſettling Abroad incorporate with other Nations, loſe their native Language, and become

come Foreigners ; but the Induſtry of thoſe that remain being diſcourag'd, the Quantity of Subſiſtence in the Country is leſſen'd, and the Support of a Family becomes more difficult. So heavy Taxes tend to diminiſh a People. 6. The Introduction of Slaves. The Negroes brought into the *Engliſh* Sugar *Iſlands*, have greatly diminiſh'd the Whites there ; the Poor are by this Means depriv'd of Employment, while a few Families acquire vaſt Eſtates ; which they ſpend on Foreign Luxuries, and educating their Children in the Habit of thoſe Luxuries ; the ſame Income is needed for the Support of one that might have maintain'd 100. The Whites who have Slaves, not labouring, are enfeebled, and therefore not ſo generally prolific ; the Slaves being work'd too hard, and ill fed, their Conſtitutions are broken, and the Deaths among them are more than the Births ; ſo that a continual Supply is needed from *Africa*. The Northern Colonies having few Slaves increaſe in Whites. Slaves alſo pejorate the Families that uſe them ; the white Chlidren become proud, diſguſted with Labour, and being educated in Idleneſs, are rendered unfit to get a Living by Induſtry.

14. Hence

14. Hence the Prince that acquires new Territory, if he finds it vacant, or removes the Natives to give his own People Room; the Legiſlator that makes effectual Laws for promoting of Trade, increaſing Employment, improving Land by more or better Tillage; providing more Food by Fiſheries; ſecuring Property, &c. and the Man that invents new Trades, Arts or Manufactures, or new Improvements in Huſbandry, may be properly called *Fathers* of their Nation, as they are the Cauſe of the Generation of Multitudes, by the Encouragement they afford to Marriage.

15. As to Privileges granted to the married, (ſuch as the *Jus trium Liberorum* among the *Romans*), they may haſten the filling of a Country that has been thinned by War or Peſtilence, or that has otherwiſe vacant Territory; but cannot increaſe a People beyond the Means provided for their Subſiſtence.

16. Foreign Luxuries & needleſs Manufactures imported and uſed in a Nation, do, by the ſame Reaſoning, increaſe the People of the Nation that furniſhes them, and diminiſh the People of the Nation that uſes them.—Laws therefore that prevent ſuch Importations, and on the contrary promote the Exportation of Manufactures to be

be conſumed in Foreign Countries, may be called (with Reſpect to the People that make them) *generative Laws*, as by increaſing Subſiſtence they encourage Marriage. Such Laws likewiſe ſtrengthen a Country, doubly, by increaſing its own People and diminiſhing its Neighbours.

17. Some *European* Nations prudently refuſe to conſume the Manufactures of *Eaſt-India* :— They ſhould likewiſe forbid them to their Colonies ; for the Gain to the Merchant, is not to be compar'd with the Loſs by this Means of People to the Nation.

18. Home Luxury in the Great, increaſes the Nation's Manufacturers employ'd by it, who are many, and only tends to diminiſh the Families that indulge in it, who are few. The greater the common faſhionable Expence of any Rank of People, the more cautious they are of Marriage. Therefore Luxury ſhould never be ſuffer'd to become common.

19. The great Increaſe of Offspring in particular Families, is not always owing to greater Fecundity of Nature, but ſometimes to Examples of Induſtry in the Heads, and induſtrious Education ; by which the Children are enabled to provide better for

themſelves,

themſelves, and their marrying early, is encouraged from the Proſpect of good Subſiſtence.

20. If there be a Sect therefore, in our Nation, that regard Frugality and Induſtry as religious Duties, and educate their Children therein, more than others commonly do ; ſuch Sect muſt conſequently increaſe more by natural Generation, than any other Sect in *Britain.*———

21. The Importation of Foreigners into a Country that has as many Inhabitants as the preſent Employments and Proviſions for Subſiſtence will bear ; will be in the End no Increaſe of People ; unleſs the New Comers have more Induſtry and Frugality than the Natives, and then they will provide more Subſiſtence, and increaſe in the Country ; but they will gradually eat the Natives out.——Nor is it neceſſary to bring in Foreigners to fill up any occaſional Vacancy in a Country ; for ſuch Vacancy (if the Laws are good, § 14, 16) will ſoon be filled by natural Generation. Who can now find the Vacancy made in *Sweden, France* or other Warlike Nations, by the Plague of Heroiſm 40 Years ago ; in *France*, by the Expulſion of the Proteſtants ; in *England*, by the Settlement of her Colonies ; or in *Guinea*, by 100 Years Exportation of Slaves,

Slaves, that has blacken'd half *America* ?—— The thinneſs of Inhabitants in *Spain*, is owing to National Pride and Idleneſs, and other Cauſes, rather than to the Expulſion of the *Moors*, or to the making of new Settlements.

22. There is in ſhort, no Bound to the prolific Nature of Plants or Animals, but what is made by their crowding and interfering with each others Means of Subſiſtence. Was the Face of the Earth vacant of other Plants, it might be gradually ſowed and overſpread with one Kind only; as, for Inſtance, with Fennel; and were it empty of other Inhabitants, it might in a few Ages be repleniſh'd from one Nation only; as, for Inſtance, with *Engliſhmen*. Thus there are ſuppos'd to be now upwards of One Million *Engliſh* Souls in *North-America*, (tho' 'tis thought ſcarce 80,000 have been brought over Sea) and yet perhaps there is not one the fewer in *Britain*, but rather many more, on Account of the Employment the Colonies afford to Manufacturers at Home. This Million doubling, ſuppoſe but once in 25 Years, will in another Century be more than the People of *England*, and the greateſt Number of *Engliſhmen* will be on this Side the Water. What an Acceſſion of Power to the *Britiſh* Empire

Empire by Sea as well as Land! What Increaſe of Trade and Navigation! What Numbers of Ships and Seamen! We have been here but little more than 100 Years, and yet the Force of our Privateers in the late War, united, was greater, both in Men and Guns, than that of the whole *Britiſh* Navy in Queen *Elizabeth*'s Time.——How important an Affair then to *Britain*, is the preſent Treaty for ſettling the Bounds between her Colonies and the *French*, and how careful ſhould ſhe be to ſecure Room enough, ſince on the Room depends ſo much the Increaſe of her People?

23. In fine, A Nation well regulated is like a Polypus; take away a Limb, itsPlace is ſoon ſupply'd; cut it in two, and each deficient Part ſhall ſpeedily grow out of the Part remaining. Thus if you have Room and Subſiſtence enough, as you may by dividing, make ten Polypes out of one, you may of one make ten Nations, equally populous and powerful; or rather, increaſe a Nation ten fold in Numbers and Strength.

And ſince Detachments of *Engliſh* from *Britain* ſent to *America*, will have their Places at Home ſo ſoon ſupply'd and increaſe ſo largely here; why ſhould the *Palatine Boors* be ſuffered to ſwarm into our Settlements, and by herding together eſtabliſh

eſtabliſh their Language and Manners to theExcluſion of ours? Why ſhould *Pennſylvania*, founded by the*Engliſh*, become a Colony of*Aliens*,who will ſhortly be ſo numerous as to Germanize us inſtead of our Anglifying them, and will never adopt our Language or Cuſtoms, any more than they ca n acquire our Complexion.

24. Which leads me to add oneRemark: That the Number of purely white People in the World is proportionably very ſmall. All *Africa* is black or tawny. *Aſia* chiefly tawny. *America* (excluſive of the new Comers) wholly ſo. And in *Europe*, the *Spaniards*, *Italians*, *French*, *Ruſſians* and *Swedes*,are generally of what we call a ſwarthyComplexion; as are the*Germans* alſo,the *Saxons* only excepted,who with the *Engliſh*, make the principal Body of White People on the Face of the Earth. I could wiſh theirNumbers were increaſed. And while we are, as I may call it, *Scouring* our Planet, by clearing *America* of Woods, and ſo making this Side of our Globe reflect a brighter Light to the Eyes of Inhabitants in *Mars* or *Venus*, why ſhould we in the Sight of Superior Beings,darken its People? why increaſe the Sons of *Africa*, by Planting them in *America*, where we have ſo fair

fair an Opportunity, by excluding all Blacks and Tawneys, of increafing the lovely White and Red? But perhaps I am partial to the Complexion of my Country, for fuch Kind of Partiality is natural to Mankind.

THE END.

15A

Reprinted from pages 1-6 of *An Essay on the Principle of Population; or, A View of Its Past and Present Effects on Human Happiness; With an Inquiry into Our Prospects Respecting the Future Removal or Mitigation of Evils Which It Occasions,* 6th ed., J. Murray, London, 1826

STATEMENT OF THE SUBJECT. RATIOS OF THE INCREASE OF POPULATION AND FOOD

T. R. Malthus

In an inquiry concerning the improvement of society, the mode of conducting the subject which naturally presents itself, is,

1. To investigate the causes that have hitherto impeded the progress of mankind towards happiness; and,

2. To examine the probability of the total or partial removal of these causes in future.

To enter fully into this question, and to enumerate all the causes that have hitherto influenced human improvement, would be much beyond the power of an individual. The principal object of the present essay is to examine the effects of one great cause intimately united with the very nature of man; which, though it has been constantly and powerfully operating since the commencement of society, has been little noticed by the writers who have treated this subject. The facts which establish the existence of this cause have, indeed, been repeatedly stated and acknowledged; but its natural and necessary effects have been almost totally overlooked; though probably among these effects may be reckoned a very considerable portion of that vice and misery, and of that unequal distribution of the bounties of nature, which it has been the unceasing object of the enlightened philanthropist in all ages to correct.

The cause to which I allude, is the constant tendency in all animated life to increase beyond the nourishment prepared for it.

It is observed by Dr. Franklin, that there is no bound to the prolific nature of plants or animals, but what is made by their crowding and interfering with each other's means of subsistence. Were the face of the earth, he says, vacant of other plants, it might be gradually sowed and overspread with one kind only, as for instance with fennel and were it empty of other inhabitants, it might in a few ages be replenished from one nation only as for instance with Englishmen.*

This is incontrovertibly true. Through the animal and vegetable kingdoms Nature has scattered the seeds of life abroad with the most pro-

* Franklin's Miscell. p. 9.

fuse and liberal hand; but has been comparatively sparing in the room and the nourishment necessary to rear them. The germs of existence contained in this earth, if they could freely develope themselves, would fill millions of worlds in the course of a few thousand years. Necessity, that imperious, all-pervading law of nature, restrains them within the prescribed bounds. The race of plants and the race of animals shrink under this great restrictive law; and man cannot by any efforts of reason escape from it.

In plants and irrational animals, the view of the subject is simple. They are all impelled by a powerful instinct to the increase of their species; and this instinct is interrupted by no doubts about providing for their offspring. Wherever therefore there is liberty, the power of increase is exerted; and the superabundant effects are repressed afterwards by want of room and nourishment.

The effects of this check on man are more complicated. Impelled to the increase of his species by an equally powerful instinct, reason interrupts his career, and asks him whether he may not bring beings into the world, for whom he cannot provide the means of support. If he attend to this natural suggestion, the restriction too frequently produces vice. If he hear it not, the human race will be constantly endeavouring to increase beyond the means of subsistence. But as, by that law of our nature which makes food necessary to the life of man, population can never actually increase beyond the lowest nourishment capable of supporting it, a strong check on population, from the difficulty of acquiring food, must be constantly in operation. This difficulty must fall somewhere, and must necessarily be severely felt in some or other of the various forms of misery, or the fear of misery, by a large portion of mankind.

That population has this constant tendency to increase beyond the means of subsistence, and that it is kept to its necessary level by these causes, will sufficiently appear from a review of the different states of society in which man has existed. But, before we proceed to this review, the subject will, perhaps, be seen in a clearer light, if we endeavour to ascertain what would be the natural increase of population, if left to exert itself with perfect freedom; and what might be expected to be the rate of increase in the productions of the earth, under the most favourable circumstances of human industry.

It will be allowed that no country has hitherto been known, where the manners were so pure and simple, and the means of subsistence so abundant, that no check whatever has existed to early marriages from the difficulty of providing for a family, and that no waste of the human species has been occasioned by vicious customs, by towns, by unhealthy occupations, or too severe labour. Consequently in no state that we have yet known, has the power of population been left to exert itself with perfect freedom.

Whether the law of marriage be instituted, or not, the dictate of nature and virtue seems to be an early attachment to one woman; and where there were no impediments of any kind in the way of an union to which such an attachment would lead, and no causes of depopulation afterwards, the increase of the human species would be evidently much greater than any increase which has been hitherto known.

In the northern states of America, where the means of subsistence have been more ample, the manners of the people more pure, and the checks to early marriages fewer, than in any of the modern states of Europe, the population has been found to double itself, for above a century and a half successively, in less than twenty-five years.* Yet, even during these periods, in some of the towns, the deaths exceeded the births,† a circumstance which clearly proves that, in those parts of the country which supplied this deficiency, the increase must have been much more rapid than the general average.

In the back settlements, where the sole employment is agriculture, and vicious customs and unwholesome occupations are little known, the population has been found to double itself in fifteen years.‡ Even this extraordinary rate of increase is probably short of the utmost power of

* It appears, from some recent calculations and estimates, that from the first settlement of America, to the year 1800, the periods of doubling have been but very little above twenty years. See a note on the increase of American population in Book ii. chap. xi

† Price's Observ. on Revers. Pay. vol. i. p. 274. 4th edit

‡ Id. p. 282.

population. Very severe labour is requisite to clear a fresh country; such situations are not in general considered as particularly healthy; and the inhabitants, probably, are occasionally subject to the incursions of the Indians, which may destroy some lives, or at any rate diminish the fruits of industry.

According to a table of Euler, calculated on a mortality of 1 in 36, if the births be to the deaths in the proportion of 3 to 1, the period of doubling will be only 12 years and 4-5ths.* And this proportion is not only a possible supposition, but has actually occurred for short periods in more countries than one.

Sir William Petty supposes a doubling possible in so short a time as ten years.†

But, to be perfectly sure that we are far within the truth, we will take the slowest of these rates of increase, a rate in which all concurring testimonies agree, and which has been repeatedly ascertained to be from procreation only.

It may safely be pronounced, therefore, that population, when unchecked, goes on doubling itself every twenty-five years, or increases in a geometrical ratio.

[*Editor's Note:* Material has been omitted at this point.]

* See this table at the end of chap. iv. book ii
† Polit. Arith. p. 14.

15B

Reprinted from pages 35–37 and 39 of *An Essay on the Principle of Population; or, A View of Its Past and Present Effects on Human Happiness; With an Inquiry into Our Prospects Respecting the Future Removal or Mitigation of Evils Which It Occasions,* 6th ed., J. Murray, London, 1826

OF THE CHECKS TO POPULATION AMONG THE AMERICAN INDIANS

T. R. Malthus

WE may next turn our view to the vast continent of America, the greatest part of which was found to be inhabited by small independent tribes of savages, subsisting, nearly like the natives of New Holland, on the productions of unassisted nature. The soil was covered by an almost universal forest, and presented few of those fruits and esculent vegetables which grow in such profusion in the islands of the South Sea. The produce of a most rude and imperfect agriculture, known to some of the tribe of hunters, was so trifling as to be considered only as a feeble aid to the subsistence acquired by the chase. The inhabitants of this new world therefore might be considered as living principally by hunting and fishing;* and the narrow limits to this mode of subsistence are obvious. The supplies derived from fishing could reach only those who were within a certain distance of the lakes, the rivers, or the sea-shore; and the ignorance and indolence of the improvi-

* Robertson's History of America, vol. ii. b. iv. p. 127. et seq. octavo edit. 1780.

dent savage would frequently prevent him from extending the benefits of these supplies much beyond the time when they were actually obtained. The great extent of territory required for the support of the hunter has been repeatedly stated and acknowledged.* The number of wild animals within his reach, combined with the facility with which they may be either killed or insnared, must necessarily limit the number of his society. The tribes of hunters, like beasts of prey, whom they resemble in their mode of subsistence, will consequently be thinly scattered over the surface of the earth. Like beasts of prey, they must either drive away or fly from every rival, and be engaged in perpetual contests with each other.†

Under such circumstances, that America should be very thinly peopled in proportion to its extent of territory, is merely an exemplification of the obvious truth, that population cannot increase without the food to support it. But the interesting part of the inquiry, that part, to which I would wish particularly to draw the attention of the reader, is, the mode by which the population is kept down to the level of this scanty supply. It cannot escape observation, that an insufficient supply of food to any people does not shew itself merely in the shape of famine, but in other more permanent forms of distress, and in generating certain customs, which operate sometimes with

* Franklin's Miscell. p. 2

† Robertson, b. iv. p. 129

greater force in the prevention of a rising population than in its subsequent destruction.

[*Editor's Note:* Material has been omitted at this point.]

In every part of the world, one of the most general characteristics of the savage is to despise and degrade the female sex.† Among most of the tribes in America their condition is so peculiarly grievous, that servitude is a name too mild to describe their wretched state. A wife is no better than a beast of burden. While the man passes his days in idleness or amusement, the woman is condemned to incessant toil. Tasks are imposed upon her without mercy, and services are received without complacence or gratitude.‡ There are some districts in America where this state of degradation has been so severely felt, that mothers have destroyed their female infants, to deliver them at once from a life in which they were doomed to such a miserable slavery.§

[*Editor's Note:* Material has been omitted at this point.]

† Robertson, b. iv. p. 103. Lettres Edif. passim. Charlevoix, Hist. Nouv. Fr. tom. iii. p. 287. Voy. de Pérouse, c. ix. p. 402, 4to. London.

‡ Robertson, b. iv. p. 105. Lettres Edif. tom. vi. p. 329. Major Roger's North America, p. 211. Creuxii Hist. Canad. p. 57.

§ Robertson, b. iv. p. 106. Raynal, Hist. des Indes, tom. iv. vii. p. 110, 8vo. 10 vol. 1795.

15C

Reprinted from pages 158–164 of *An Essay on the Principle of Population; or, A View of Its Past and Present Effects on Human Happiness; With an Inquiry into Our Prospects Respecting the Future Removal of Mitigation of Evils Which It Occasions*, 6th ed., J. Murray, London, 1826

OF THE CHECKS TO POPULATION IN DIFFERENT PARTS OF AFRICA

T. R. Malthus

[*Editor's Note:* In the original, material precedes this excerpt]

The description, which Bruce gives of some parts of the country which he passed through on his return home, presents a picture more dreadful even than the state of Abyssinia, and shews how little population depends on the birth of children, in comparison of the production of food and those circumstances of natural and political situation which influence this produce.

" At half past six," Bruce says, " we arrived " at Garigana, a village whose inhabitants had all

* Bruce, vol. ii. p. 559.

"perished with hunger the year before; their "wretched bones being all unburied and scat-"tered upon the surface of the ground where the "village formerly stood. We encamped among "the bones of the dead; no space could be found "free from them."*

Of another town or village in his route he observes, "The strength of Teawa was 25 horse. "The rest of the inhabitants might be 1200 "naked miserable and despicable Arabs, like "the rest of those which live in villages.**** Such "was the state of Teawa. Its consequence was "only to remain till the Daveina Arabs should "resolve to attack it, when its corn-fields being "burnt and destroyed in a night by a multitude "of horsemen, the bones of its inhabitants scat-"tered upon the earth would be all its remains, "like those of the miserable village of Gari-"gana."†

"There is no water between Teawa and Beyla. "Once Indedidema and a number of villages "were supplied with water from wells, and had "large crops of Indian corn sown about their "possessions. The curse of that country, the "Daveina Arabs, have destroyed Indedidema "and all the villages about it; filled up their "wells, burnt their crops, and exposed all the in-"habitants to die by famine."‡

* Bruce, vol. iv. p. 349.
† Id. p. 353.
‡ Id. p. 411

Soon after leaving Sennaar, he says, "We "began to see the effects of the quantity of rain "having failed. There was little corn sown, and "that so late as to be scarcely above ground. It "seems the rains begin later as they pass north-"ward. Many people were here employed in "gathering grass-seeds to make a very bad kind "of bread. These people appear perfect skele-"tons, and no wonder, as they live upon such "fare. Nothing increases the danger of travel-"ling and prejudice against strangers more, than "the scarcity of provisions in the country through "which you are to pass."*

"Came to Eltic, a straggling village about half "a mile from the Nile, in the north of a large "bare plain; all pasture, except the banks of the "river which are covered with wood. We now "no longer saw any corn sown. The people "here were at the same miserable employment "as those we had seen before, that of gathering "grass-seeds."†

Under such circumstances of climate and political situation, though a greater degree of foresight, industry and security, might considerably better their condition and increase their population, the birth of a greater number of children without these concomitants would only aggravate their misery, and leave their population where it was.

* Bruce, vol. iv. p. 511.
† Id. p. 511.

The same may be said of the once flourishing and populous country of Egypt. Its present depressed state has not been caused by the weakening of the principle of increase, but by the weakening of the principle of industry and foresight, from the insecurity of property consequent on a most tyrannical and oppressive government. The principle of increase in Egypt at present does all that is possible for it to do. It keeps the population fully up to the level of the means of subsistence; and, were its power ten times greater than it really is, it could do no more.

The remains of ancient works, the vast lakes, canals and large conduits for water destined to keep the Nile under control, serving as reservoirs to supply a dry year, and as drains and outlets to prevent the superabundance of water in wet years, sufficiently indicate to us that the former inhabitants of Egypt by art and industry contrived to fertilize a much greater quantity of land from the overflowings of their river, than is done at present; and to prevent, in some measure, the distresses which are now so frequently experienced from a redundant or insufficient inundation.* It is said of the governor Petronius, that, effecting by art what was denied by nature, he caused abundance to prevail in Egypt under the disadvantages of such a deficient inundation, as had always before been accompanied by dearth.† A flood too

* Bruce, vol. iii. c. xvii. p. 710.

† Voyage de Volney, tom. i. c. iii. p. 33. 8vo.

great is as fatal to the husbandman as one that is deficient; and the ancients had, in consequence drains and outlets to spread the superfluous waters over the thirsty sands of Lybia, and render even the desert habitable. These works are now all out of repair, and by ill management often produce mischief instead of good. The causes of this neglect, and consequently of the diminished means of subsistence, are obviously to be traced to the extreme ignorance and brutality of the government, and the wretched state of the people. The Mamelukes, in whom the principal power resides, think only of enriching themselves, and employ for this purpose what appears to them to be the simplest method, that of seizing wealth wherever it may be found, of wresting it by violence from the possessor, and of continually imposing new and arbitrary contributions.* Their ignorance and brutality, and the constant state of alarm in which they live, prevent them from having any views of enriching the country, the better to prepare it for their plunder. No public works therefore are to be expected from the government, and no individual proprietor dares to undertake any improvement which might imply the possession of capital, as it would probably be the immediate signal of his destruction. Under such circumstances we cannot be surprised that the ancient works are neglected, that the soil is ill cultivated, and that the means of subsistence, and conse-

* Voyage de Volney, tom. i. c. xii. p. 170.

quently the population, are greatly reduced. But such is the natural fertility of the Delta from the inundations of the Nile, that even without any capital employed upon the land, without a right of succession, and consequently almost without a right of property, it still maintains a considerable population in proportion to its extent, sufficient, if property were secure, and industry well directed, gradually to improve and extend the cultivation of the country and restore it to its former state of prosperity. It may be safely pronounced of Egypt that it is not the want of population that has checked its industry, but the want of industry that has checked its population.

The immediate causes which keep down the population to the level of the present contracted means of subsistence, are but too obvious. The peasants are allowed for their maintenance only sufficient to keep them alive.* A miserable sort of bread made of doura without leaven or flavour, cold water, and raw onions make up the whole of their diet. Meat and fat, of which they are passionately fond, never appear but on great occasions, and among those who are more at their ease. Their habitations are huts made of earth, where a stranger would be suffocated with the heat and smoke; and where the diseases generated by want of cleanliness, by moisture and by bad nourishment, often visit them and commit great ravages. To these physical evils are added

* Voyage de Volney, tom. i. c. xii. p. 172.

a constant state of alarm, the fear of the plunder of the Arabs, and the visits of the Mamelukes, the spirit of revenge transmitted in families, and all the evils of a continual civil war.*

In the year 1783 the plague was very fatal; and in 1784 and 1785 a dreadful famine reigned in Egypt, owing to a deficiency in the inundation of the Nile. Volney draws a frightful picture of the misery that was suffered on this occasion. The streets of Cairo, which at first were full of beggars, were soon cleared of all these objects, who either perished or fled. A vast number of unfortunate wretches, in order to escape death, spread themselves over all the neighbouring countries, and the towns of Syria were inundated with Egyptians. The streets and public places were crowded by famished and dying skeletons. All the most revolting modes of satisfying the cravings of hunger were resorted to; the most disgusting food was devoured with eagerness; and Volney mentions the having seen under the walls of ancient Alexandria two miserable wretches seated on the carcase of a camel, and disputing with the dogs its putrid flesh. The depopulation of the two years was estimated at one-sixth of all the inhabitants.†

* Volney, tom. i. c. xii. p. 173. This sketch of the state of the peasantry in Egypt given by Volney seems to be nearly confirmed by all other writers on this subject; and particularly in a valuable paper entitled *Considérations générales sur l'Agriculture de l'Egypte, par L. Reynier.* (Mémoires sur l' Egypte, tom. iv. p. 1.)

† Voy. de Volney, tom. i. c. xii. s. ii.

16

Reprinted from *Linn. Soc. London (Zool.) J.* **3**:45–62 (1858)

On the Tendency of Species to form Varieties; and on the Perpetuation of Varieties and Species by Natural Means of Selection. By CHARLES DARWIN, Esq., F.R.S., F.L.S., & F.G.S., and ALFRED WALLACE, Esq. Communicated by Sir CHARLES LYELL, F.R.S., F.L.S., and J. D. HOOKER, Esq., M.D., V.P.R.S., F.L.S., &c.

[Read July 1st, 1858.]

London, June 30th, 1858.

MY DEAR SIR,—The accompanying papers, which we have the honour of communicating to the Linnean Society, and which all relate to the same subject, viz. the Laws which affect the Production of Varieties, Races, and Species, contain the results of the investigations of two indefatigable naturalists, Mr. Charles Darwin and Mr. Alfred Wallace.

These gentlemen having, independently and unknown to one another, conceived the same very ingenious theory to account for the appearance and perpetuation of varieties and of specific forms on our planet, may both fairly claim the merit of being original thinkers in this important line of inquiry; but neither of them having published his views, though Mr. Darwin has for many years past been repeatedly urged by us to do so, and both authors having now unreservedly placed their papers in our hands, we think it would best promote the interests of science that a selection from them should be laid before the Linnean Society.

Taken in the order of their dates, they consist of:—

1. Extracts from a MS. work on Species*, by Mr. Darwin, which was sketched in 1839, and copied in 1844, when the copy was read by Dr. Hooker, and its contents afterwards communicated to Sir Charles Lyell. The first Part is devoted to "The Variation of Organic Beings under Domestication and in their Natural State;" and the second chapter of that Part, from which we propose to read to the Society the extracts referred to, is headed, "On the Variation of Organic Beings in a state of Nature; on the Natural Means of Selection; on the Comparison of Domestic Races and true Species."

2. An abstract of a private letter addressed to Professor Asa Gray, of Boston, U.S., in October 1857, by Mr. Darwin, in which

* This MS. work was never intended for publication, and therefore was not written with care.—C. D. 1858.

he repeats his views, and which shows that these remained unaltered from 1839 to 1857.

3. An Essay by Mr. Wallace, entitled "On the Tendency of Varieties to depart indefinitely from the Original Type." This was written at Ternate in February 1858, for the perusal of his friend and correspondent Mr. Darwin, and sent to him with the expressed wish that it should be forwarded to Sir Charles Lyell, if Mr. Darwin thought it sufficiently novel and interesting. So highly did Mr. Darwin appreciate the value of the views therein set forth, that he proposed, in a letter to Sir Charles Lyell, to obtain Mr. Wallace's consent to allow the Essay to be published as soon as possible. Of this step we highly approved, provided Mr. Darwin did not withhold from the public, as he was strongly inclined to do (in favour of Mr. Wallace), the memoir which he had himself written on the same subject, and which, as before stated, one of us had perused in 1844, and the contents of which we had both of us been privy to for many years. On representing this to Mr. Darwin, he gave us permission to make what use we thought proper of his memoir, &c.; and in adopting our present course, of presenting it to the Linnean Society, we have explained to him that we are not solely considering the relative claims to priority of himself and his friend, but the interests of science generally; for we feel it to be desirable that views founded on a wide deduction from facts, and matured by years of reflection, should constitute at once a goal from which others may start, and that, while the scientific world is waiting for the appearance of Mr. Darwin's complete work, some of the leading results of his labours, as well as those of his able correspondent, should together be laid before the public.

We have the honour to be yours very obediently,

CHARLES LYELL.
JOS. D. HOOKER.

J. J. Bennett, Esq.,
Secretary of the Linnean Society.

I. *Extract from an unpublished Work on Species, by* C. DARWIN, Esq., *consisting of a portion of a Chapter entitled, "On the Variation of Organic Beings in a state of Nature; on the Natural Means of Selection; on the Comparison of Domestic Races and true Species."*

De Candolle, in an eloquent passage, has declared that all nature is at war, one organism with another, or with external nature.

Seeing the contented face of nature, this may at first well be doubted; but reflection will inevitably prove it to be true. The war, however, is not constant, but recurrent in a slight degree at short periods, and more severely at occasional more distant periods; and hence its effects are easily overlooked. It is the doctrine of Malthus applied in most cases with tenfold force. As in every climate there are seasons, for each of its inhabitants, of greater and less abundance, so all annually breed; and the moral restraint which in some small degree checks the increase of mankind is entirely lost. Even slow-breeding mankind has doubled in twenty-five years; and if he could increase his food with greater ease, he would double in less time. But for animals without artificial means, the amount of food for each species must, *on an average*, be constant, whereas the increase of all organisms tends to be geometrical, and in a vast majority of cases at an enormous ratio. Suppose in a certain spot there are eight pairs of birds, and that *only* four pairs of them annually (including double hatches) rear only four young, and that these go on rearing their young at the same rate, then at the end of seven years (a short life, excluding violent deaths, for any bird) there will be 2048 birds, instead of the original sixteen. As this increase is quite impossible, we must conclude either that birds do not rear nearly half their young, or that the average life of a bird is, from accident, not nearly seven years. Both checks probably concur. The same kind of calculation applied to all plants and animals affords results more or less striking, but in very few instances more striking than in man.

Many practical illustrations of this rapid tendency to increase are on record, among which, during peculiar seasons, are the extraordinary numbers of certain animals; for instance, during the years 1826 to 1828, in La Plata, when from drought some millions of cattle perished, the whole country actually *swarmed* with mice. Now I think it cannot be doubted that during the breeding-season all the mice (with the exception of a few males or females in excess) ordinarily pair, and therefore that this astounding increase during three years must be attributed to a greater number than usual surviving the first year, and then breeding, and so on till the third year, when their numbers were brought down to their usual limits on the return of wet weather. Where man has introduced plants and animals into a new and favourable country, there are many accounts in how surprisingly few years the whole country has become stocked with them. This increase would necessarily stop as soon as the country was fully stocked; and yet we have every reason to believe, from what is known of wild animals, that *all* would pair in the spring. In the majority of cases it is most difficult to imagine where the checks fall—though generally, no doubt, on the seeds, eggs, and young; but when we remember how impossible, even in mankind (so much better known than any other animal), it is to infer from repeated casual observations what the average duration of life is, or to discover the different percentage of deaths to births in different countries, we ought to feel no surprise at our being unable to discover where the check falls in any animal or plant. It should always be remembered, that in most cases the checks are recurrent yearly in a small, regular degree, and in an extreme degree during unusually cold, hot, dry, or wet years, according to the constitution of the being in question. Lighten any check in the least degree, and the geometrical powers of increase in every organism will almost instantly increase the average number of the favoured species. Nature may be compared to a surface on which rest ten thousand sharp wedges touching each other and driven inwards by incessant blows. Fully to realize these views much reflection is requisite. Malthus on man should be studied; and all such cases as those of the mice in La Plata, of the cattle and horses when first turned out in South America, of the birds by our calculation, &c., should be well considered. Reflect on the enormous multiplying power *inherent and annually in action* in all animals; reflect on the countless seeds scattered by a hundred ingenious contrivances year after year, over the whole face of the land; and yet we have every reason to suppose that the average percentage of each of the inhabitants of a country usually remains constant. Finally let it be borne in mind that this average number of individuals (the external conditions remaining the same) in each country is kept up by recurrent struggles against other species or against external nature (as on the borders of the Arctic regions, where the cold checks life), and that ordinarily each individual of every species holds its place, either by its own struggle and capacity of acquiring nourishment in some period of its life, from the egg upwards; or by the struggle of its parents (in short-lived organisms, when the main check occurs at longer intervals) with other individuals of the *same* or *different* species.

But let the external conditions of a country alter. If in a small degree, the relative proportions of the inhabitants will in most cases simply be slightly changed; but let the number of inhabitants be small, as on an island, and free access to it from other countries be circumscribed, and let the change of conditions continue progressing (forming new stations), in such a case the original inhabitants must cease to be as perfectly adapted to the changed conditions as they were originally. It has been shown in a former part of this work, that such changes of external conditions would, from their acting on the reproductive system, probably cause the organization of those beings which were most affected to become, as under domestication, plastic. Now, can it be doubted, from the struggle each individual has to obtain subsistence, that any minute variation in structure, habits, or instincts, adapting that individual better to the new conditions, would tell upon its vigour and health? In the struggle it would have a better *chance* of surviving; and those of its offspring which inherited the variation, be it ever so slight, would also have a better *chance*. Yearly more are bred than can survive; the smallest grain in the balance, in the long run, must tell on which death shall fall, and which shall survive. Let this work of selection on the one hand, and death on the other, go on for a thousand generations, who will pretend to affirm that it would produce no effect, when we remember what, in a few years, Bakewell effected in cattle, and Western in sheep, by this identical principle of selection?

To give an imaginary example from changes in progress on an island:—let the organization of a canine animal which preyed chiefly on rabbits, but sometimes on hares, become slightly plastic; let these same changes cause the number of rabbits very slowly to decrease, and the number of hares to increase; the effect of this would be that the fox or dog would be driven to try to catch more hares: his organization, however, being slightly plastic, those individuals with the lightest forms, longest limbs, and best eyesight, let the difference be ever so small, would be slightly favoured, and would tend to live longer, and to survive during that time of the year when food was scarcest; they would also rear more young, which would tend to inherit these slight peculiarities. The less fleet ones would be rigidly destroyed. I can see no more reason to doubt that these causes in a thousand generations would produce a marked effect, and adapt the form of the fox or dog to the catching of hares instead of rabbits, than that greyhounds can be improved by selection and careful breeding. So would it be with plants under similar circumstances. If the number of individuals of a species with plumed seeds could be increased by greater powers of dissemination within its own area (that is, if the check to increase fell chiefly on the seeds), those seeds which were provided with ever so little more down, would in the long run be most disseminated; hence a greater number of seeds thus formed would germinate, and would tend to produce plants inheriting the slightly better-adapted down*.

Besides this natural means of selection, by which those individuals are preserved, whether in their egg, or larval, or mature state, which are best adapted to the place they fill in nature, there is a second agency at work in most unisexual animals, tending to produce the same effect, namely, the struggle of the males for the females. These struggles are generally decided by the law of battle, but in the case of birds, apparently, by the charms of their song, by their beauty or their power of courtship, as in the dancing rock-thrush of Guiana. The most vigorous and healthy males, implying perfect adaptation, must generally gain the victory in their contests. This kind of selection, however, is less rigorous than the other; it does not require the death of the less successful, but gives to them fewer descendants. The struggle falls, moreover, at a time of year when food is generally abundant, and perhaps the effect chiefly produced would be the modification of the secondary sexual characters, which are not related to the power of obtaining food, or to defence from enemies, but to fighting with or rivalling other males. The result of this struggle amongst the males may be compared in some respects to that produced by those agriculturists who pay less attention to the careful selection of all their young animals, and more to the occasional use of a choice mate.

II. *Abstract of a Letter from* C. Darwin, Esq., *to* Prof. Asa Gray, *Boston, U.S., dated Down, September 5th,* 1857.

1. It is wonderful what the principle of selection by man, that is the picking out of individuals with any desired quality, and breeding from them, and again picking out, can do. Even breeders have been astounded at their own results. They can act on differences inappreciable to an uneducated eye. Selection has been *methodically* followed in *Europe* for only the last half century; but it was occasionally, and even in some degree methodically, followed in the most ancient times. There must have been also a kind of unconscious selection from a remote period, namely in

* I can see no more difficulty in this, than in the planter improving his varieties of the cotton plant.—C. D. 1858.

the preservation of the individual animals (without any thought of their offspring) most useful to each race of man in his particular circumstances. The "roguing," as nurserymen call the destroying of varieties which depart from their type, is a kind of selection. I am convinced that intentional and occasional selection has been the main agent in the production of our domestic races; but however this may be, its great power of modification has been indisputably shown in later times. Selection acts only by the accumulation of slight or greater variations, caused by external conditions, or by the mere fact that in generation the child is not absolutely similar to its parent. Man, by this power of accumulating variations, adapts living beings to his wants—may be said to make the wool of one sheep good for carpets, of another for cloth, &c.

2. Now suppose there were a being who did not judge by mere external appearances, but who could study the whole internal organization, who was never capricious, and should go on selecting for one object during millions of generations; who will say what he might not effect? In nature we have some *slight* variation occasionally in all parts; and I think it can be shown that changed conditions of existence is the main cause of the child not exactly resembling its parents; and in nature geology shows us what changes have taken place, and are taking place. We have almost unlimited time; no one but a practical geologist can fully appreciate this. Think of the Glacial period, during the whole of which the same species at least of shells have existed; there must have been during this period millions on millions of generations.

3. I think it can be shown that there is such an unerring power at work in *Natural Selection* (the title of my book), which selects exclusively for the good of each organic being. The elder De Candolle, W. Herbert, and Lyell have written excellently on the struggle for life; but even they have not written strongly enough. Reflect that every being (even the elephant) breeds at such a rate, that in a few years, or at most a few centuries, the surface of the earth would not hold the progeny of one pair. I have found it hard constantly to bear in mind that the increase of every single species is checked during some part of its life, or during some shortly recurrent generation. Only a few of those annually born can live to propagate their kind. What a trifling difference must often determine which shall survive, and which perish!

4. Now take the case of a country undergoing some change. This will tend to cause some of its inhabitants to vary slightly—

not but that I believe most beings vary at all times enough for selection to act on them. Some of its inhabitants will be exterminated; and the remainder will be exposed to the mutual action of a different set of inhabitants, which I believe to be far more important to the life of each being than mere climate. Considering the infinitely various methods which living beings follow to obtain food by struggling with other organisms, to escape danger at various times of life, to have their eggs or seeds disseminated, &c. &c., I cannot doubt that during millions of generations individuals of a species will be occasionally born with some slight variation, profitable to some part of their economy. Such individuals will have a better chance of surviving, and of propagating their new and slightly different structure; and the modification may be slowly increased by the accumulative action of natural selection to any profitable extent. The variety thus formed will either coexist with, or, more commonly, will exterminate its parent form. An organic being, like the woodpecker or misseltoe, may thus come to be adapted to a score of contingences—natural selection accumulating those slight variations in all parts of its structure, which are in any way useful to it during any part of its life.

5. Multiform difficulties will occur to every one, with respect to this theory. Many can, I think, be satisfactorily answered. *Natura non facit saltum* answers some of the most obvious. The slowness of the change, and only a very few individuals undergoing change at any one time, answers others. The extreme imperfection of our geological records answers others.

6. Another principle, which may be called the principle of divergence, plays, I believe, an important part in the origin of species. The same spot will support more life if occupied by very diverse forms. We see this in the many generic forms in a square yard of turf, and in the plants or insects on any little uniform islet, belonging almost invariably to as many genera and families as species. We can understand the meaning of this fact amongst the higher animals, whose habits we understand. We know that it has been experimentally shown that a plot of land will yield a greater weight if sown with several species and genera of grasses, than if sown with only two or three species. Now, every organic being, by propagating so rapidly, may be said to be striving its utmost to increase in numbers. So it will be with the offspring of any species after it has become diversified into varieties, or subspecies, or true species. And it follows, I think, from the foregoing facts, that the varying offspring of each species will try

(only few will succeed) to seize on as many and as diverse places in the economy of nature as possible. Each new variety or species, when formed, will generally take the place of, and thus exterminate its less well-fitted parent. This I believe to be the origin of the classification and affinities of organic beings at all times; for organic beings always *seem* to branch and sub-branch like the limbs of a tree from a common trunk, the flourishing and diverging twigs destroying the less vigorous—the dead and lost branches rudely representing extinct genera and families.

This sketch is *most* imperfect; but in so short a space I cannot make it better. Your imagination must fill up very wide blanks.

C. DARWIN.

III. *On the Tendency of Varieties to depart indefinitely from the Original Type.* By ALFRED RUSSEL WALLACE.

One of the strongest arguments which have been adduced to prove the original and permanent distinctness of species is, that *varieties* produced in a state of domesticity are more or less unstable, and often have a tendency, if left to themselves, to return to the normal form of the parent species; and this instability is considered to be a distinctive peculiarity of all varieties, even of those occurring among wild animals in a state of nature, and to constitute a provision for preserving unchanged the originally created distinct species.

In the absence or scarcity of facts and observations as to *varieties* occurring among wild animals, this argument has had great weight with naturalists, and has led to a very general and somewhat prejudiced belief in the stability of species. Equally general, however, is the belief in what are called "permanent or true varieties,"—races of animals which continually propagate their like, but which differ so slightly (although constantly) from some other race, that the one is considered to be a *variety* of the other. Which is the *variety* and which the original *species*, there is generally no means of determining, except in those rare cases in which the one race has been known to produce an offspring unlike itself and resembling the other. This, however, would seem quite incompatible with the "permanent invariability of species," but the difficulty is overcome by assuming that such varieties have strict limits, and can never again vary further from the original type, although they may return to it, which, from the analogy of the domesticated animals, is considered to be highly probable, if not certainly proved.

It will be observed that this argument rests entirely on the assumption, that *varieties* occurring in a state of nature are in all respects analogous to or even identical with those of domestic animals, and are governed by the same laws as regards their permanence or further variation. But it is the object of the present paper to show that this assumption is altogether false, that there is a general principle in nature which will cause many *varieties* to survive the parent species, and to give rise to successive variations departing further and further from the original type, and which also produces, in domesticated animals, the tendency of varieties to return to the parent form.

The life of wild animals is a struggle for existence. The full exertion of all their faculties and all their energies is required to preserve their own existence and provide for that of their infant offspring. The possibility of procuring food during the least favourable seasons, and of escaping the attacks of their most dangerous enemies, are the primary conditions which determine the existence both of individuals and of entire species. These conditions will also determine the population of a species; and by a careful consideration of all the circumstances we may be enabled to comprehend, and in some degree to explain, what at first sight appears so inexplicable—the excessive abundance of some species, while others closely allied to them are very rare.

The general proportion that must obtain between certain groups of animals is readily seen. Large animals cannot be so abundant as small ones; the carnivora must be less numerous than the herbivora; eagles and lions can never be so plentiful as pigeons and antelopes; the wild asses of the Tartarian deserts cannot equal in numbers the horses of the more luxuriant prairies and pampas of America. The greater or less fecundity of an animal is often considered to be one of the chief causes of its abundance or scarcity; but a consideration of the facts will show us that it really has little or nothing to do with the matter. Even the least prolific of animals would increase rapidly if unchecked, whereas it is evident that the animal population of the globe must be stationary, or perhaps, through the influence of man, decreasing. Fluctuations there may be; but permanent increase, except in restricted localities, is almost impossible. For example, our own observation must convince us that birds do not go on increasing every year in a geometrical ratio, as they would do, were there not some powerful check to their natural increase. Very few birds produce less than two young ones each year, while many have six, eight, or ten; four will certainly be below the average; and if we suppose that each pair produce young only four times in their life, that will also be below the average, supposing them not to die either by violence or want of food. Yet at this rate how tremendous would be the increase in a few years from a single pair! A simple calculation will show that in fifteen years each pair of birds would have increased to nearly ten millions! whereas we have no reason to believe that the number of the birds of any country increases at all in fifteen or in one hundred and fifty years. With such powers of increase the population must have reached its limits, and have become stationary, in a very few years after the origin of each species. It is evident, therefore, that each year an immense number of birds must perish—as many in fact as are born; and as on the lowest calculation the progeny are each year twice as numerous as their parents, it follows that, whatever be the average number of individuals existing in any given country, *twice that number must perish annually*,—a striking result, but one which seems at least highly probable, and is perhaps under rather than over the truth. It would therefore appear that, as far as the continuance of the species and the keeping up the average number of individuals are concerned, large broods are superfluous. On the average all above *one* become food for hawks and kites, wild cats and weasels, or perish of cold and hunger as winter comes on. This is strikingly proved by the case of particular species; for we find that their abundance in individuals bears no relation whatever to their fertility in producing offspring. Perhaps the most remarkable instance of an immense bird population is that of the passenger pigeon of the United States, which lays only one, or at most two eggs, and is said to rear generally but one young one. Why is this bird so extraordinarily abundant, while others producing two or three times as many young are much less plentiful? The explanation is not difficult. The food most congenial to this species, and on which it thrives best, is abundantly distributed over a very extensive region, offering such differences of soil and climate, that in one part or another of the area the supply never fails. The bird is capable of a very rapid and long-continued flight, so that it can pass without fatigue over the whole of the district it inhabits, and as soon as the supply of food begins to fail in one place is able to discover a fresh feeding-ground. This example strikingly shows us that the procuring a constant supply of wholesome food is almost the sole condition requisite for ensuring the rapid increase of a given species, since neither the limited fecundity, nor the unrestrained attacks of birds of prey and of man are here sufficient to check it. In no other birds are these peculiar circumstances so strikingly combined. Either their food is more liable to failure, or they have not sufficient power of wing to search for it over an extensive area, or during some season of the year it becomes very scarce, and less wholesome substitutes have to be found; and thus, though more fertile in offspring, they can never increase beyond the supply of food in the least favourable seasons. Many birds can only exist by migrating, when their food becomes scarce, to regions possessing a milder, or at least a different climate, though, as these migrating birds are seldom excessively abundant, it is evident that the countries they visit are still deficient in a constant and abundant supply of wholesome food. Those whose organization does not permit them to migrate when their food becomes periodically scarce, can never attain a large population. This is probably the reason why woodpeckers are scarce with us, while in the tropics they are among the most abundant of solitary birds. Thus the house sparrow is more abundant than the redbreast, because its food is more constant and plentiful,—seeds of grasses being preserved during the winter, and our farm-yards and stubble-fields furnishing an almost inexhaustible supply. Why, as a general rule, are aquatic, and especially sea birds, very numerous in individuals? Not because they are more prolific than others, generally the contrary; but because their food never fails, the sea-shores and river-banks daily swarming with a fresh supply of small mollusca and crustacea. Exactly the same laws will apply to mammals. Wild cats are prolific and have few enemies; why then are they never as abundant as rabbits? The only intelligible answer is, that their supply of food is more precarious. It appears evident, therefore, that so long as a country remains physically unchanged, the numbers of its animal population cannot materially increase. If one species does so, some others requiring the same kind of food must diminish in proportion. The numbers that die annually must be immense; and as the individual existence of each animal depends upon itself, those that die must be the weakest—the very young, the aged, and the diseased,—while those that prolong their existence can only be the most perfect in health and vigour—those who are best able to obtain food regularly, and avoid their numerous enemies. It is, as we commenced by remarking, "a struggle for existence," in which the weakest and least perfectly organized must always succumb.

Now it is clear that what takes place among the individuals of a species must also occur among the several allied species of a group,—viz. that those which are best adapted to obtain a regular supply of food, and to defend themselves against the attacks of their enemies and the vicissitudes of the seasons, must necessarily obtain and preserve a superiority in population; while those species which from some defect of power or organization are the least capable of counteracting the vicissitudes of food, supply, &c., must diminish in numbers, and, in extreme cases, become altogether extinct. Between these extremes the species will present various degrees of capacity for ensuring the means of preserving life; and it is thus we account for the abundance or rarity of species. Our ignorance will generally prevent us from accurately tracing the effects to their causes; but could we become perfectly acquainted with the organization and habits of the various species of animals, and could we measure the capacity of each for performing the different acts necessary to its safety and existence under all the varying circumstances by which it is surrounded, we might be able even to calculate the proportionate abundance of individuals which is the necessary result.

If now we have succeeded in establishing these two points—1st, *that the animal population of a country is generally stationary, being kept down by a periodical deficiency of food, and other checks*; and, 2nd, *that the comparative abundance or scarcity of the individuals of the several species is entirely due to their organization and resulting habits, which, rendering it more difficult to procure a regular supply of food and to provide for their personal safety in some cases than in others, can only be balanced by a difference in the population which have to exist in a given area*—we shall be in a condition to proceed to the consideration of *varieties*, to which the preceding remarks have a direct and very important application.

Most or perhaps all the variations from the typical form of a species must have some definite effect, however slight, on the habits or capacities of the individuals. Even a change of colour might, by rendering them more or less distinguishable, affect their safety; a greater or less development of hair might modify their habits. More important changes, such as an increase in the power or dimensions of the limbs or any of the external organs, would more or less affect their mode of procuring food or the range of country which they inhabit. It is also evident that most changes would affect, either favourably or adversely, the powers of prolonging existence. An antelope with shorter or weaker legs must necessarily suffer more from the attacks of the feline carnivora; the passenger pigeon with less powerful wings would sooner or later be affected in its powers of procuring a regular supply of food; and in both cases the result must necessarily be a diminution of the population of the modified species. If, on the other hand, any species should produce a variety having slightly increased powers of preserving existence, that variety must inevitably in time acquire a superiority in numbers. These results must follow as surely as old age, intemperance, or scarcity of food produce an increased mortality. In both cases there may be many individual exceptions; but on the average the rule will invariably be found to hold good. All varieties will therefore fall into two classes—those which under the same conditions would never reach the population of the parent species, and those which would in time obtain and keep a numerical superiority. Now, let some alteration of physical conditions occur in the district—a long period of drought, a destruction of vegetation by locusts, the irruption of some new carnivorous animal seeking "pastures new"—any change in fact tending to render existence more difficult to the species in question, and tasking its utmost powers to avoid complete extermination; it is evident that, of all the individuals composing the species, those forming the least numerous and most feebly organized variety would suffer first, and, were the pressure severe, must soon become extinct. The same causes continuing in action, the parent species would next suffer, would gradually diminish in numbers, and with a recurrence of similar unfavourable conditions might also become extinct. The superior variety would then alone remain, and on a return to favourable circumstances would rapidly increase in numbers and occupy the place of the extinct species and variety.

The *variety* would now have replaced the *species*, of which it would be a more perfectly developed and more highly organized form. It would be in all respects better adapted to secure its safety, and to prolong its individual existence and that of the race. Such a variety *could not* return to the original form; for that form is an inferior one, and could never compete with it for existence. Granted, therefore, a "tendency" to reproduce the original type of the species, still the variety must ever remain preponderant in numbers, and under adverse physical conditions *again alone survive*.

But this new, improved, and populous race might itself, in course of time, give rise to new varieties, exhibiting several diverging modifications of form, any of which, tending to increase the facilities for preserving existence, must, by the same general law, in their turn become predominant. Here, then, we have *progression and continued divergence* deduced from the general laws which regulate the existence of animals in a state of nature, and from the undisputed fact that varieties do frequently occur. It is not, however, contended that this result would be invariable; a change of physical conditions in the district might at times materially modify it, rendering the race which had been the most capable of supporting existence under the former conditions now the least so, and even causing the extinction of the newer and, for a time, superior race, while the old or parent species and its first inferior varieties continued to flourish. Variations in unimportant parts might also occur, having no perceptible effect on the life-preserving powers; and the varieties so furnished might run a course parallel with the parent species, either giving rise to further variations or returning to the former type. All we argue for is, that certain varieties have a tendency to maintain their existence longer than the original species, and this tendency must make itself felt; for though the doctrine of chances or averages can never be trusted to on a limited scale, yet, if applied to high numbers, the results come nearer to what theory demands, and, as we approach to an infinity of examples, become strictly accurate. Now the scale on which nature works is so vast—the numbers of individuals and periods of time with which she deals approach so near to infinity, that any cause, however slight, and however liable to be veiled and counteracted by accidental circumstances, must in the end produce its full legitimate results.

Let us now turn to domesticated animals, and inquire how varieties produced among them are affected by the principles here enunciated. The essential difference in the condition of wild and domestic animals is this,—that among the former, their well-being and very existence depend upon the full exercise and healthy condition of all their senses and physical powers, whereas, among the latter, these are only partially exercised, and in some cases are absolutely unused. A wild animal has to search, and often to labour, for every mouthful of food—to exercise sight, hearing, and smell in seeking it, and in avoiding dangers, in procuring shelter from the inclemency of the seasons, and in providing for the subsistence and safety of its offspring. There is no muscle of its body that is not called into daily and hourly activity; there is no sense or faculty that is not strengthened by continual exercise. The domestic animal, on the other hand, has food provided for it, is sheltered, and often confined, to guard it against the vicissitudes of the seasons, is carefully secured from the attacks of its natural enemies, and seldom even rears its young without human assistance. Half of its senses and faculties are quite useless; and the other half are but occasionally called into feeble exercise, while even its muscular system is only irregularly called into action.

Now when a variety of such an animal occurs, having increased power or capacity in any organ or sense, such increase is totally useless, is never called into action, and may even exist without the animal ever becoming aware of it. In the wild animal, on the contrary, all its faculties and powers being brought into full action for the necessities of existence, any increase becomes immediately available, is strengthened by exercise, and must even slightly modify the food, the habits, and the whole economy of the race. It creates as it were a new animal, one of superior powers, and which will necessarily increase in numbers and outlive those inferior to it.

Again, in the domesticated animal all variations have an equal chance of continuance; and those which would decidedly render a wild animal unable to compete with its fellows and continue its existence are no disadvantage whatever in a state of domesticity. Our quickly fattening pigs, short-legged sheep, pouter pigeons, and poodle dogs could never have come into existence in a state of nature, because the very first step towards such inferior forms would have led to the rapid extinction of the race; still less could they now exist in competition with their wild allies. The great speed but slight endurance of the race horse, the unwieldy strength of the ploughman's team, would both be useless in a state of nature. If turned wild on the pampas, such animals would probably soon become extinct, or under favourable circumstances might each lose those extreme qualities which would never be called into action, and in a few generations would revert to a common type, which must be that in which the various powers and faculties are so proportioned to each other as to be best adapted to procure food and secure safety,—that in which by the full exercise of every part of his organization the animal can alone continue to live. Domestic varieties, when turned wild, *must* return to something near the type of the original wild stock, *or become altogether extinct.*

We see, then, that no inferences as to varieties in a state of nature can be deduced from the observation of those occurring among domestic animals. The two are so much opposed to each other in every circumstance of their existence, that what applies to the one is almost sure not to apply to the other. Domestic animals are abnormal, irregular, artificial; they are subject to varieties which never occur and never can occur in a state of nature: their very existence depends altogether on human care; so far are many of them removed from that just proportion of faculties, that true balance of organization, by means of which alone an animal left to its own resources can preserve its existence and continue its race.

The hypothesis of Lamarck—that progressive changes in species have been produced by the attempts of animals to increase the development of their own organs, and thus modify their structure and habits—has been repeatedly and easily refuted by all writers on the subject of varieties and species, and it seems to have been considered that when this was done the whole question has been finally settled; but the view here developed renders such an hypothesis quite unnecessary, by showing that similar results must be produced by the action of principles constantly at work in nature. The powerful retractile talons of the falcon- and the cat-tribes have not been produced or increased by the volition of those animals; but among the different varieties which occurred in the earlier and less highly organized forms of these groups, *those always survived longest which had the greatest facilities for seizing their prey.* Neither did the giraffe acquire its long neck by desiring to reach the foliage of the more lofty shrubs, and constantly stretching its neck for the purpose, but because any varieties which occurred among its antitypes with a longer neck than usual *at once secured a fresh range of pasture over the same ground as their shorter-necked companions, and on the first scarcity of food were thereby enabled to outlive them.* Even the peculiar colours of many animals, especially insects, so closely resembling the soil or the leaves or the trunks on which they habitually reside, are explained on the same principle; for though in the course of ages varieties of many tints may have occurred, *yet those races having colours best adapted to concealment from their enemies would inevitably survive the longest.* We have also here an acting cause to account for that balance so often observed in nature,—a deficiency in one set of organs always being compensated by an increased development of some others—powerful wings accompanying weak feet, or great velocity making up for the absence of defensive weapons; for it has been shown that all varieties in which an unbalanced deficiency occurred could not long continue their existence. The action of this principle is exactly like that of the centrifugal governor of the steam engine, which checks and corrects any irregularities almost before they become evident; and in like manner no unbalanced deficiency in the animal kingdom can ever reach any conspicuous magnitude, because it would make itself felt at the very first step, by rendering existence difficult and extinction almost sure soon to follow. An origin such as is here advocated will also agree with the peculiar character of the modifications of form and structure which obtain in organized beings—the many lines of divergence from a central type, the increasing efficiency and power of a particular organ through a succession of allied species, and the remarkable persistence of unimportant parts such as colour, texture of plumage and hair, form of horns or crests, through a series of species differing considerably in more essential characters. It also furnishes us with a reason for that "more specialized structure" which Professor Owen states to be a characteristic of recent compared with extinct forms, and which would evidently be the result of the progressive modification of any organ applied to a special purpose in the animal economy.

We believe we have now shown that there is a tendency in nature to the continued progression of certain classes of *varieties* further and further from the original type—a progression to which there appears no reason to assign any definite limits—and that the same principle which produces this result in a state of nature will also explain why domestic varieties have a tendency to revert to the original type. This progression, by minute steps, in various directions, but always checked and balanced by the necessary conditions, subject to which alone existence can be preserved, may, it is believed, be followed out so as to agree with all the phenomena presented by organized beings, their extinction and succession in past ages, and all the extraordinary modifications of form, instinct, and habits which they exhibit.

Ternate, February, 1858.

Part VI

NATURAL SELECTION THEORY IN THE ORIGIN OF SPECIES: 1859–1872

Editor's Comments on Papers 17 Through 20

In all six editions of his *On the Origin of Species by Means of Natural Selection, or the Preservation of Favoured Races in the Struggle for Life* (1859–1872, excerpts from the first edition are reprinted as Papers 17A, 17B, 17C, and 17D) Charles Darwin contended that three types of selection were involved in bringing about "descent with modification" (evolution).* He divided the ecological interactions an organism has that cause evolution into the following three categories: "man's power of selection" (artificial selection, see Bajema, 1982), natural selection, and sexual selection (Fig. 1). By

*All the changes in text that Charles Darwin made in the second and later editions are catalogued by Morse Peckham, editor, in *The Origin of Species by Charles Darwin, A Variorum Text* (Philadelphia: University of Pennsylvania Press, 1959), 816p.

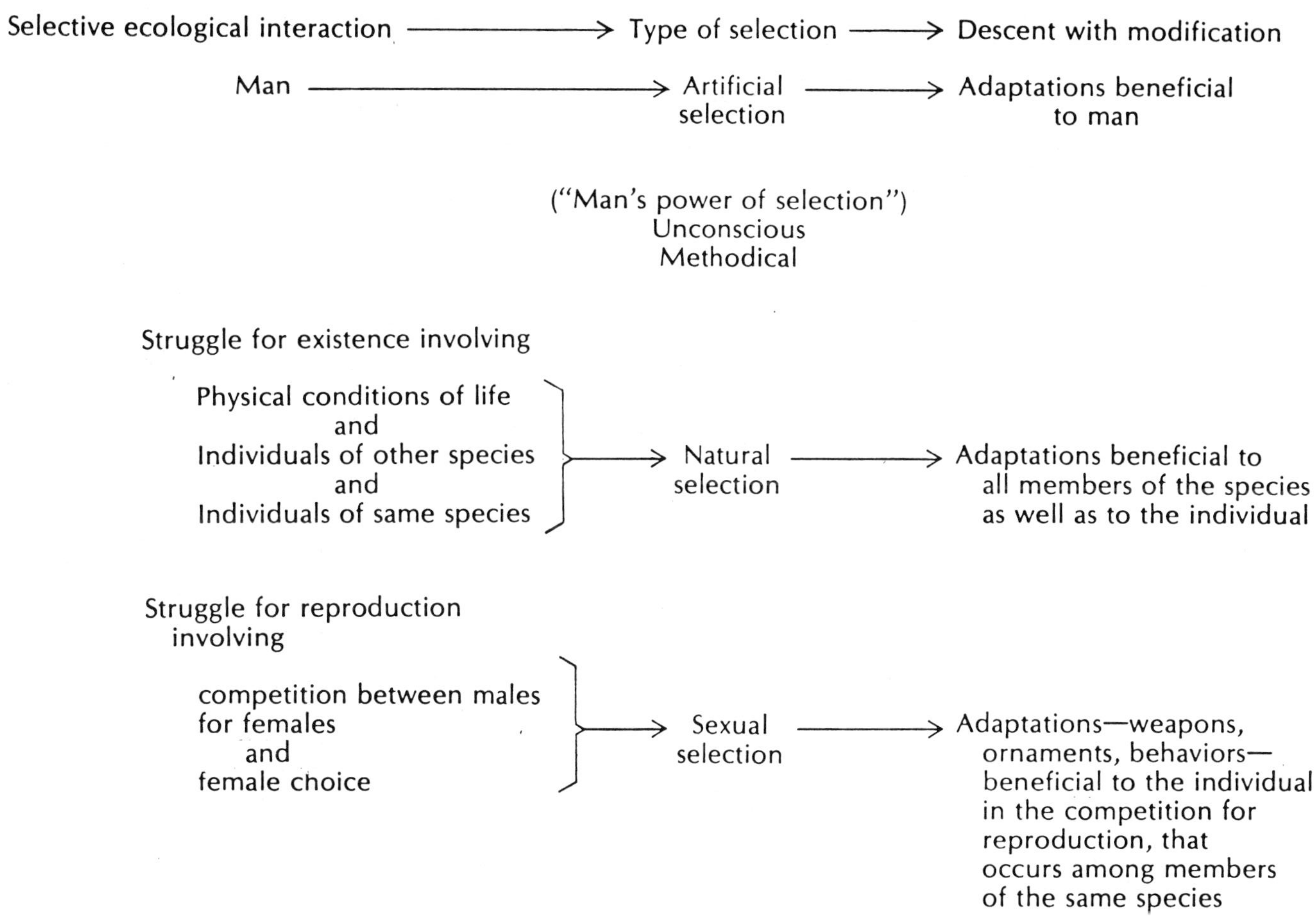

Figure 1. Charles Darwin's 1859 system for classifying selection. (After G. Hardin and C. Bajema, *Biology: Its Principles and Implications,* 3rd ed. San Francisco: W. H. Freeman, 1978.)

dividing selection into these three categories, Darwin was able, first, to argue by analogy from artificial selection—a process already widely known to bring about hereditary changes in domesticated species—to both natural and sexual selection, which were operating in the rest of nature (Figs. 2 and 3). Second, the division of selection into those three categories enabled him to partition those intraspecific ecological interactions that selected for adaptations that were advantageous to the individuals possessing them but harmful to certain other members of the species into a separate selective system—sexual selection.

Papers 17A-D reprint Darwin's 1859 arguments by analogy from artificial selection to both natural selection (17A, p. 61; 17B, pp. 83-84; 17D, p. 469) and sexual selection (17B, pp. 88-89). Darwin emphasized that: "I have called this principle, by which each slight variation, if useful, is preserved, by the term of Natural Selection, in order to mark its relation to man's power of selection" (Paper 17A, p. 61).

Darwin also deduced "descent with modification" by both natural and sexual selection from the overproduction of offspring, the struggle for existence and for mates, and the presence of heritable variations with respect to both the ability to survive and to obtain mates (Paper 17A, pp. 62-64; 17D, pp. 467-468; see also Figs. 2 and 3). He pointed out that:

> I should premise that I use the term Struggle for Existence in a large and metaphorical sense, including dependence of one being on another, and including (which is more important) not only the life of the individual, but success in leaving progeny. (Paper 17A, p. 62)

The historian Michael Ruse (1971, pp. 316-317) contends that Darwin's "struggle for existence" would more properly be construed as the "struggle for reproduction" were it not for the fact that Darwin divided selection operating in nature into natural selection and sexual selection. The struggle for existence involved in natural selection "is basically a struggle to survive long enough to *reproduce,* that is, to leave offspring" (Ruse, 1971, p. 317). Sexual selection involves the struggle for reproduction among those who have survived.

Darwin considered the possibility that, in addition to natural and sexual selection, there might be other natural causes producing "descent with modification." He came to the conclusion that the "direct action of the external conditions of life" (climate, food, and so on) was an unimportant cause of individual variations that were adaptive (Darwin, 1859, pp. 3, 10, 29, 132). While Darwin rejected what he considered to be the Lamarckian theory that individual

THE LOGIC OF CHARLES DARWIN'S THEORY OF NATURAL SELECTION

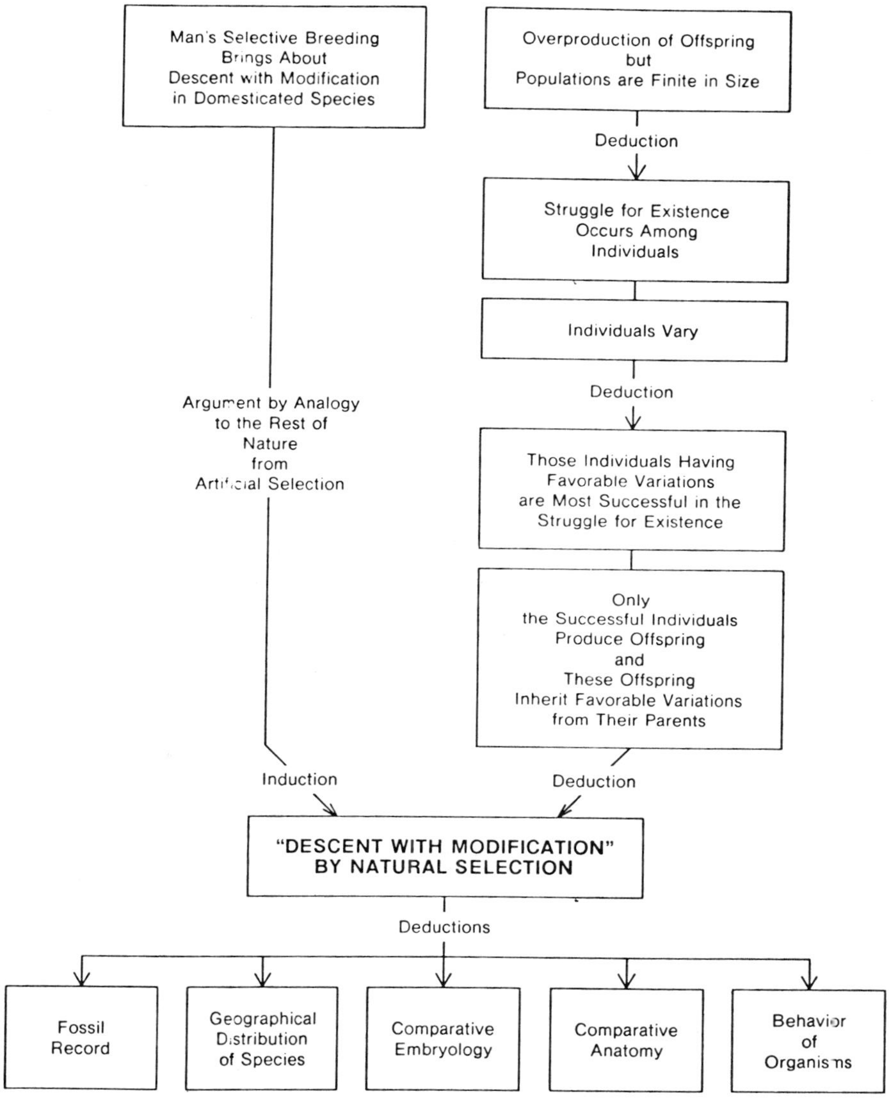

Figure 2. Darwin used three lines of reasoning in developing and advocating his theory of natural selection: the inductive vera causa ("true cause") argument from artificial selection; the deduction of the set of causes that produce natural selection; and the rationalist vera causa used to deduce the kinds of observations that would be made in nature if natural selection operated the way that Darwin predicted. (After M. Ruse, *The Darwinian Revolution: Science Red Tooth and Claw,* Chicago: University of Chicago Press, 1979, p. 198.)

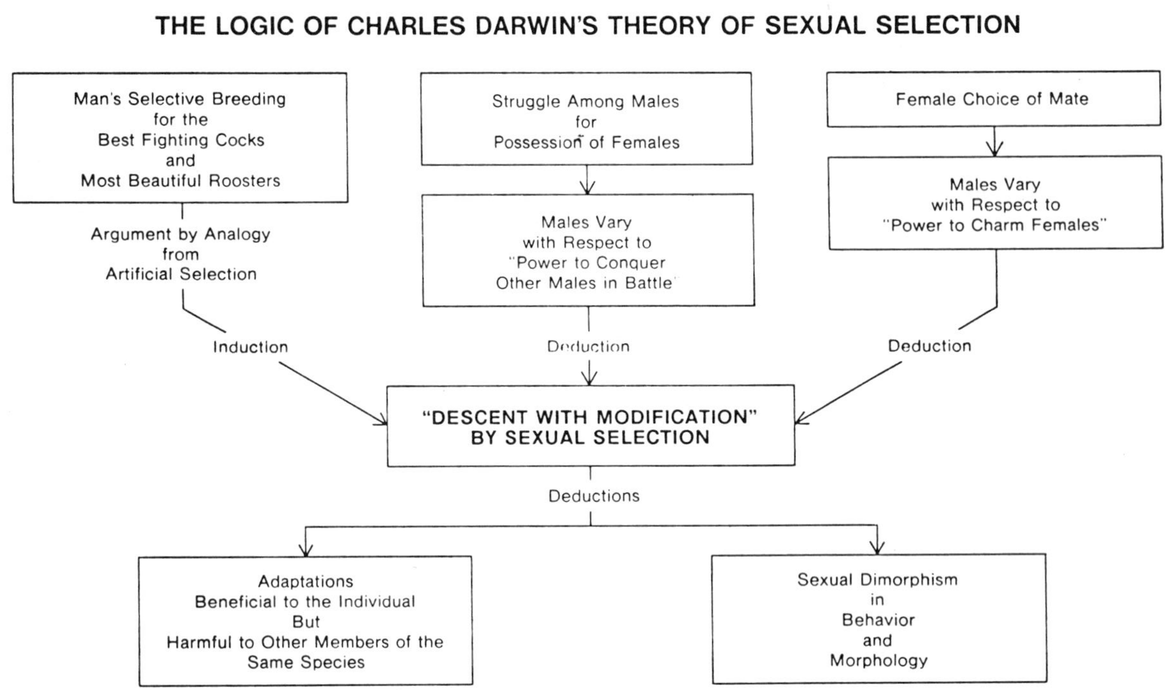

Figure 3. Darwin also used three lines of reasoning in developing and advocating his theory of sexual selection: the inductive vera causa ("true cause") argument from artificial selection; the deduction of the set of natural causes that produce sexual selection; and the rationalist vera causa used to deduce the kinds of observations that would be made in nature if sexual selection operated the way that Darwin predicted. (After M. Ruse, *The Darwinian Revolution: Science Red Tooth and Claw*, Chicago: University of Chicago Press, 1979, p. 198.)

heritable adaptations could be produced by "willing" or "volition" (Darwin, 1859, pp. 3, 242), he did accept the theory:

> that habit, use, and disuse, have, in some cases, played a considerable part in the modification of the constitution, and of the structure of various organs; but that the effects of use and disuse have often been largely combined with, and sometimes overmastered by, the natural selection of innate differences. (Darwin, 1859, p. 143)

The following challenge was issued by Darwin to his opponents concerning the kind of individual adaptations that his selection theory predicted would not be found in nature:

> Natural selection cannot possibly produce any modification in any one species exclusively for the good of another species; though throughout nature one species incessantly takes advantage of, and profits by, the structure of another. . . . If it could be proved that any part of the structure of any one species had been formed for the exclusive good of another species, it would annihilate my theory, for such could not have been produced through natural selection. (Paper 17C, pp. 200–201)

Darwin also argued that natural selection does not work to produce absolute perfection because: "Natural selection tends only to make each organic being as perfect as, or slightly more perfect than the other inhabitants of the same country with which it has to struggle for existence" (Paper 17C, p. 201).

His *Origin of Species* was, according to Darwin, one long argument from beginning to end (Barlow, 1958, p. 140). Since Darwin thought that natural selection operated so slowly that it would be very difficult to directly measure its effects in nature, he wrote in an 1861 letter that:

> I believe in nat. Selection, not because I can prove in any single case that it has changed one species into another, but because it groups & explains well (as it seems to me) a host of facts in classification, embryology, morphology, rudimentary organs, geological succession & Distribution. (in de Beer, 1958, p. 113)

Many of those who opposed Darwin's theory of natural selection admitted that selection did operate in nature but only to maintain the type of the species (Wilberforce, 1860, pp. 232–238; Tristam, 1866; and others). They denied that selection could raise the level of perfection above the normal level of a species not by denying the existence or even the potential power of selection, but by denying that "natural variations from the specific type favourable to the individual from which nature is to select can anywhere be found" (Wilberforce, 1860, p. 239). Several scholars distinguished between selection that operated to produce adaptations and selection that operated to maintain adaptations (Wilberforce, 1860; Spencer, Paper 19; Tristam, 1866, p. 120; Pearson, Paper 22).

Many readers misunderstood natural selection to mean that a

conscious agent was involved in producing adaptation. Darwin wrote in an August, 1860, letter to W. H. Harvey:

> The term "selection", I see, deceives many persons, though I see no more reason why it should than elective affinity, as used by the old chemists. If I had to rewrite my book, I would use "natural preservation" or "naturally preserved." (in Darwin and Seward, 1903, I, p. 163)

However, Darwin defended his use of natural selection as a metaphor in the third (1863) edition of the *Origin:*

> Others have objected that the term selection implies conscious choice in the animals which become modified; and it has even been urged that as plants have no volition, natural selection is not applicable to them! In the literal sense of the word, no doubt, natural selection is a misnomer[*]; but who ever objected to chemists speaking of the elective affinities of the various elements? . . . and yet an acid cannot strictly be said to elect the base with which it will in preference combine. It has been said that I speak of natural selection as an active power or Deity; but who objects to an author speaking of the attraction of gravity as ruling the movements of the planets? Every one knows what is meant and is implied by such metaphorical expressions; and they are almost necessary for brevity. So again it is difficult to avoid personifying the word Nature; but I mean by Nature, only the aggregate action and product of many natural laws, and by laws the sequence of events as ascertained by us. With a little familiarity such superficial objections will be forgotten. (in Peckham, 1959, p. 165)

The English philosopher Herbert Spencer (1820-1903) in his 1864 two volume work, *The Principles of Biology* (excerpt as Paper 18), coined the immortal metaphor "survival of the fittest" to describe in mechanical terms the operation of natural selection. He pointed out that natural selection maintains as well as produces adaptations and that it is to Darwin "we owe the discovery that natural selection is capable of *producing* fitness between organsisms and their circumstances" (Paper 18, p. 446).

In an attempt to help others better understand Darwin's and his theory concerning the natural processes that produce "descent with modification," Alfred R. Wallace wrote a letter to Charles Darwin in 1866 urging him to adopt Spencer's "survival of the fittest" metaphor to describe natural selection (Paper 19). The letter that Darwin wrote in response to Wallace is also reprinted as Paper 19. Darwin first used "survival of the fittest" to describe natural selection in his 1868 two-volume publication, *The Variation of Animals and Plants Under Domestication:*

> This preservation, during the battle of life, of varieties which possess any advantage in structure, constitution, or instinct, I have called Natural Selection; and Mr. Herbert Spencer has well

*Darwin substituted the phrase "false term" for "misnomer" in the fifth and sixth editions of the *Origin* (Peckham, 1959, p. 165).

expressed the same idea by the Survival of the Fittest. (Darwin, 1868, I, p. 6)

Darwin divided selection into natural selection (selection by death) and sexual selection (selection operating via differences among those who survive). When one considers the fact that he never used "survival of the fittest" to describe sexual selection and that natural selection as defined by Darwin operated almost exclusively by mortality, one can better understand why he adopted Spencer's metaphor to describe natural selection. However, the confusion brought about by the use of Spencer's incomplete and inaccurate metaphor "survival of the fittest"to describe natural selection can be avoided by explicitly defining the process of natural selection in terms of both its ecological causes and genetic consequences. Selection affects the reproductive success of organisms and is brought about by the ecological interactions an organism has with the physical conditions of its environment, individuals of other species, and members of its own species. The genetic consequences of selection are measured in terms of the effect such ecological interactions have on the differential reproduction of genes—evolution.

In 1862 the English naturalist Henry Bates (1825–1892) employed natural selection theory to explain the geographical distribution of mimicry in certain species of South American butterflies in "Contributions to an Insect Fauna of the Amazon Valley, *Lepidoptera: Heliconidae*" (excerpted as Paper 20). Bates observed that races of a butterfly species that mimicked another more abundant species of butterfly only occurred where populations of the mimic and mimicked species coexisted. This observation led Bates to conclude that natural selection has favored the evolution of mimics because those accidently produced varieties gain protection from potential predators by mimicking another distasteful butterfly species avoided by predators.

In an anonymous review of Bates's monograph, Darwin wrote: "It is hardly an exaggeration to say, whilst reading and reflecting on the various facts given in this Memoir, we feel to be as near witnesses, as we can ever hope to be, of the creation of a new species on this earth" (Darwin, 1863, p. 223). Darwin incorporated a discussion of mimicry into later editions of the *Origin,* summarizing Bates's observations and conclusions:

> From facts of this nature, Mr. Bates concludes that the Leptalis first varies; and when a variety happens to resemble in some degree any common butterfly inhabiting the same district, this variety, from its resemblance to a flourishing and little-persecuted kind, has a better chance of escaping destruction from predaceous birds and insects, and is consequently oftener preserved; . . . "the less perfect degrees of resemblance being generation after generation eliminated, and only the others left to propagate their

> kind." So that here we have an excellent illustration of natural selection. (Darwin, 1866, in Peckham, 1959, pp. 666–668)

It is impossible in one volume to adequately review all of the difficulties Darwin encountered or all the advances he made with respect to natural selection theory. Readers are referred to Ghiselin (1969), Glick (1973), Hull (1973), Ruse (1979) and Vorzimmer (1970) for comprehensive historical analyses of the fate of Charles Darwin's theory of "descent with modification" by natural selection from 1859 to 1882.

In all six editions of the *Origin* Charles Darwin classified sexual selection as a process distinct from but similar to natural selection. During the 1860s and 1870s Darwin assigned an increasingly important role to sexual selection as a cause of evolution, devoting over one-half of his *Descent of Man and Selection in Relation to Sex* (Darwin, 1871) to a discussion of sexual selection. The development of Darwin's ideas concerning sexual selection have been summarized by Ghiselin (1969, 1974), Kottler (1980), Ruse (1979) and Vorzimmer (1970). A companion Benchmark volume on the subject is also being prepared by this volume editor.

REFERENCES

Bajema, C. J., ed., 1982, *Artificial Selection and the Development of Evolutionary Theory,* Benchmark Papers in Systematic and Evolutionary Biology, vol. 4, Hutchinson Ross, Stroudsburg, Pa., 384p.

Barlow, N., ed., 1958, *The Autobiography of Charles Darwin,* Harcourt, Brace and World, New York, 253p. (Original omissions restored, appendix and notes added.)

Darwin, C. R., 1859, *On the Origin of Species by Means of Natural Selection or the Preservation of Favoured Races in the Struggle for Life,* 1st ed., J. Murray, London, 502p. (Facsimile edition by Harvard University Press, 1966.)

Darwin, C. R., 1863, A Review of "Contributions to an Insect Fauna of the Amazon Valley" by Henry Walter Bates, Esq., *Nat. Hist. Rev. Quart. J. Biol. Sci.,* pp. 219-224.

Darwin, C. R., 1868, *The Variation of Animals and Plants Under Domestication,* 2 vols., 1st ed., J. Murray, London, 411p., 486p.

Darwin, C. R., 1871, *The Descent of Man and Selection in Relation to Sex,* 2 vols., J. Murray, London, 423p., 475p.

Darwin, F., and A. C. Seward, eds., 1903, *More Letters of Charles Darwin,* 2 vols., Appleton, New York, 494p., 508p.

de Beer, G., 1958, Further Unpublished Letters of Charles Darwin, *Ann. Sci.* **14:**83–115.

Ghiselin, M. T., 1969, *The Triumph of the Darwinian Method,* University of California Press, Berkeley, 287p.

Ghiselin, M. T., 1974, *The Economy of Nature and the Evolution of Sex,* University of California Press, Berkeley, 346p.

Glick, T. F., ed., 1973, *The Comparative Reception of Darwinism,* University of Texas Press, Austin, 505p.

Hull, D., ed., 1973, *Darwin and His Critics: The Reception of Darwin's Theory of Evolution by the Scientific Community,* Harvard University Press, Cambridge, 473p.
Kottler, M. J., 1980, Darwin, Wallace, and the Origin of Sexual Dimorphism, *Am. Philos. Soc. Proc.* **124:**203-226.
Peckham, M., ed., 1959, *The Origin of Species by Charles Darwin, A Variorum Text,* University of Pennsylvania Press, Philadelphia, 816p.
Ruse, M., 1971, Natural Selection in *The Origin of Species, Hist. Philos. Sci.* **1:**311-351.
Ruse, M., 1979, *The Darwinian Revolution: Science Red in Tooth and Claw,* University of Chicago Press, Chicago, 320p.
Spencer, H., 1864, *Principles of Biology,* 2 vols., Williams and Norgate, London, 492p., 566p.
Tristam, H. B., 1866, Recent Geographical and Historical Progress in Zoology, *Contemp. Rev.* **2:**103-125.
Vorzimmer, P. J., 1970, *Charles Darwin: The Years of Controversy, The Origin of Species and Its Critics 1859-1882,* Temple University Press, Philadelphia, 300p.
Wilberforce, S., 1860, [A review of] *On the Origin of Species . . .* by Charles Darwin, *Quart. Rev.* **108:**225-264.

17A

Reprinted from pages 60-64 of *On the Origin of Species by Means of Natural Selection, or the Preservation of Favoured Races in the Struggle for Life,* J. Murray, London, 1859

STRUGGLE FOR EXISTENCE

C. R. Darwin

Bears on natural selection—The term used in a wide sense—Geometrical powers of increase — Rapid increase of naturalised animals and plants—Nature of the checks to increase—Competition universal — Effects of climate — Protection from the number of individuals—Complex relations of all animals and plants throughout nature—Struggle for life most severe between individuals and varieties of the same species ; often severe between species of the same genus—The relation of organism to organism the most important of all relations.

Before entering on the subject of this chapter, I must make a few preliminary remarks, to show how the struggle for existence bears on Natural Selection. It has been seen in the last chapter that amongst organic beings in a state of nature there is some individual variability ; indeed I am not aware that this has ever been disputed. It is immaterial for us whether a multitude of doubtful forms be called species or sub-species or varieties ; what rank, for instance, the two or three hundred doubtful forms of British plants are entitled to hold, if the existence of any well-marked varieties be admitted. But the mere existence of individual variability and of some few well-marked varieties, though necessary as the foundation for the work, helps us but little in understanding how species arise in nature. How have all those exquisite adaptations of one part of the organisation to another part, and to the conditions of life, and of one distinct organic being to another being, been perfected? We see these beautiful co-adaptations most plainly in the woodpecker and missletoe; and only a little less plainly in the humblest parasite which clings to the hairs of a quadruped or feathers of a bird; in the structure of the beetle which dives through the water; in the plumed seed which is wafted by the gentlest breeze; in short, we see beautiful adaptations everywhere and in every part of the organic world.

Again, it may be asked, how is it that varieties, which I have called incipient species, become ultimately converted into good and distinct species, which in most cases obviously differ from each other far more than do the varieties of the same species? How do those groups of species, which constitute what are called distinct genera, and which differ from each other more than do the species of the same genus, arise? All these results, as we shall more fully see in the next chapter, follow inevitably from the struggle for life. Owing to this struggle for life, any variation, however slight and from whatever cause proceeding, if it be in any degree profitable to an individual of any species, in its infinitely complex relations to other organic beings and to external nature, will tend to the preservation of that individual, and will generally be inherited by its offspring. The offspring, also, will thus have a better chance of surviving, for, of the many individuals of any species which are periodically born, but a small number can survive. I have called this principle, by which each slight variation, if useful, is preserved, by the term of Natural Selection, in order to mark its relation to man's power of selection. We have seen that man by selection can certainly produce great results, and can adapt organic beings to his own uses, through the accumulation of slight but useful variations, given to him by the hand of Nature. But Natural Selection, as we shall hereafter see, is a power incessantly ready for action, and is as immeasurably superior to man's feeble efforts, as the works of Nature are to those of Art.

We will now discuss in a little more detail the struggle for existence. In my future work this subject shall be treated, as it well deserves, at much greater length. The elder De Candolle and Lyell have largely and philosophically shown that all organic beings are exposed to severe competition. In regard to plants, no one has treated this subject with more spirit and ability than W. Herbert, Dean of Manchester, evidently the result of his great horticultural knowledge. Nothing is easier than to admit in words the truth of the universal struggle for life, or more difficult—at least I have found it so—than constantly to bear this conclusion in mind. Yet unless it be thoroughly engrained in the mind, I am convinced that the whole economy of nature, with every fact on distribution, rarity, abundance, extinction, and variation, will be dimly seen or quite misunderstood. We behold the face of nature bright with gladness, we often see superabundance of food; we do not see, or we forget, that the birds which are idly singing round us mostly live on insects or seeds, and are thus constantly destroying life; or we forget how largely these songsters, or their eggs, or their nestlings, are destroyed by birds and beasts of prey; we do not always bear in mind, that though food may be now superabundant, it is not so at all seasons of each recurring year.

I should premise that I use the term Struggle for Existence in a large and metaphorical sense, including dependence of one being on another, and including (which is more important) not only the life of the individual, but success in leaving progeny. Two canine animals in a time of dearth, may be truly said to struggle with each other which shall get food and live. But a plant on the edge of a desert is said to struggle for life against the drought, though more properly it should be said to be dependent on the moisture. A plant which annually produces a thousand seeds, of which on an average only one comes to maturity, may be more truly said to struggle with the plants of the same and other kinds which already clothe the ground. The missletoe is dependent on the apple and a few other trees, but can only in a far-fetched sense be said to struggle with these trees, for if too many of these parasites grow on the same tree, it will languish and die. But several seedling missletoes, growing close together on the same branch, may more truly be said to struggle with each other. As the missletoe is disseminated by birds, its existence depends on birds; and it may metaphorically be said to struggle with other fruit-bearing plants, in order to tempt birds to devour and thus disseminate its seeds rather than those of other plants. In these several senses, which pass into each other, I use for convenience sake the general term of struggle for existence.

A struggle for existence inevitably follows from the high rate at which all organic beings tend to increase. Every being, which during its natural lifetime produces several eggs or seeds, must suffer destruction during some period of its life, and during some season or occasional year, otherwise, on the principle of geometrical increase, its numbers would quickly become so inordinately great that no country could support the product. Hence, as more individuals are produced than can possibly survive, there must in every case be a struggle for existence, either one individual with another of the same species, or with the individuals of distinct species, or with the physical conditions of life. It is the doctrine of Malthus applied with manifold force to the whole animal and vegetable kingdoms; for in this case there can be no artificial increase of food, and no prudential restraint from marriage. Although some species may be now increasing, more or less rapidly, in numbers, all cannot do so, for the world would not hold them.

There is no exception to the rule that every organic being naturally increases at so high a rate, that if not destroyed, the earth would soon be covered by the progeny of a single pair. Even slow-breeding man has doubled in twenty-five years, and at this rate, in a few thousand years, there would literally not be standing room for his progeny. Linnæus has calculated that if an annual plant produced only two seeds—and there is no plant so unproductive as this—and their seedlings next year produced two, and so on, then in twenty years there would be a million plants. The elephant is reckoned to be the slowest breeder of all known animals, and I have taken some pains to estimate its probable minimum rate of natural increase: it will be under the mark to assume that it breeds when thirty years old, and goes on breeding till ninety years old, bringing forth three pair of young in this interval; if this be so, at the end of the fifth century there would be alive fifteen million elephants, descended from the first pair.

But we have better evidence on this subject than mere theoretical calculations, namely, the numerous recorded cases of the astonishingly rapid increase of various animals in a state of nature, when circumstances have been favourable to them during two or three following seasons. Still more striking is the evidence from our domestic animals of many kinds which have run wild in several parts of the world: if the statements of the rate of increase of slow-breeding cattle and horses in South-America, and latterly in Australia, had not been well authenticated, they would have been quite incredible. So it is with plants: cases could be given of introduced plants which have become common throughout whole islands in a period of less than ten years.

[*Editor's Note:* Material has been omitted at this point.]

17B

Reprinted from pages 80–90 and 107–109 of *On the Origin of Species by Means of Natural Selection, or the Preservation of Favoured Races in the Struggle for Life,* J. Murray, London, 1859

NATURAL SELECTION

C. R. Darwin

Natural Selection—its power compared with man's selection—its power on characters of trifling importance—its power at all ages and on both sexes—Sexual Selection—On the generality of intercrosses between individuals of the same species—Circumstances favourable and unfavourable to Natural Selection, namely, intercrossing, isolation, number of individuals—Slow action—Extinction caused by Natural Selection—Divergence of Character, related to the diversity of inhabitants of any small area, and to naturalisation—Action of Natural Selection, through Divergence of Character and Extinction, on the descendants from a common parent—Explains the Grouping of all organic beings.

How will the struggle for existence, discussed too briefly in the last chapter, act in regard to variation? Can the principle of selection, which we have seen is so potent in the hands of man, apply in nature? I think we shall see that it can act most effectually. Let it be borne in mind in what an endless number of strange peculiarities our domestic productions, and, in a lesser degree, those under nature, vary; and how strong the hereditary tendency is. Under domestication, it may be truly said that the whole organisation becomes in some degree plastic. Let it be borne in mind how infinitely complex and close-fitting are the mutual relations of all organic beings to each other and to their physical conditions of life. Can it, then, be thought improbable, seeing that variations useful to man have undoubtedly occurred, that other variations useful in some way to each being in the great and complex battle of life, should sometimes occur in the course of thousands of generations? If such do occur, can we doubt (remembering that many more individuals are born than can possibly survive) that individuals having any advantage, however slight, over others, would have the best chance of surviving and of procreating their kind? On the other hand, we may feel sure that any variation in the least degree injurious would be rigidly destroyed. This preservation of favourable variations and the rejection of injurious variations, I call Natural Selection. Variations neither useful nor injurious would not be affected by natural selection, and would be left a fluctuating element, as perhaps we see in the species called polymorphic.

We shall best understand the probable course of natural selection by taking the case of a country undergoing some physical change, for instance, of climate. The proportional numbers of its inhabitants would almost immediately undergo a change, and some species might become extinct. We may conclude, from what we have seen of the intimate and complex manner in which the inhabitants of each country are bound together, that any change in the numerical proportions of some of the inhabitants, independently of the change of climate itself, would most seriously affect many of the others. If the country were open on its borders, new forms would certainly immigrate, and this also would seriously disturb the relations of some of the former inhabitants. Let it be remembered how powerful the influence of a single introduced tree or mammal has been shown to be. But in the case of an island, or of a country partly surrounded by barriers, into which new and better adapted forms could not freely enter, we should then have places in the economy of nature which would assuredly be better filled up, if some of the original inhabitants were in some manner modified; for, had the area been open to immigration, these same places would have been seized on by intruders. In such case, every slight modification, which in the course of ages chanced to arise, and which in any way favoured the individuals of any of the species, by better adapting them to their altered conditions, would tend to be preserved; and natural selection would thus have free scope for the work of improvement.

We have reason to believe, as stated in the first chapter, that a change in the conditions of life, by specially acting on the reproductive system, causes or increases variability; and in the foregoing case the conditions of life are supposed to have undergone a change, and this would manifestly be favourable to natural selection, by giving a better chance of profitable variations occurring; and unless profitable variations do occur, natural selection can do nothing. Not that, as I believe, any extreme amount of variability is necessary; as man can certainly produce great results by adding up in any given direction mere individual differences, so could Nature, but far more easily, from having incomparably longer time at her disposal. Nor do I believe that any great physical change, as of climate, or any unusual degree of isolation to check immigration, is actually necessary to produce new and unoccupied places for natural selection to fill up by modifying and improving some of the varying inhabitants. For as all the inhabitants of each country are struggling together with nicely balanced forces, extremely slight modifications in the structure or habits of one inhabitant would often give it an advantage over others; and still further modifications of the same kind would often still further increase the advantage. No country can be named in which all the native inhabitants are now so perfectly adapted to each other and to the physical conditions under which they live, that none of them could anyhow be improved; for in all countries, the natives have been so far conquered by naturalised productions, that they have allowed foreigners to take firm possession of the land. And as foreigners have thus everywhere beaten some of the natives, we may safely conclude that the natives might have been modified with advantage, so as to have better resisted such intruders.

As man can produce and certainly has produced a great result by his methodical and unconscious means of selection, what may not nature effect? Man can act only on external and visible characters: nature cares nothing for appearances, except in so far as they may be useful to any being. She can act on every internal organ, on every shade of constitutional difference, on the whole machinery of life. Man selects only for his own good; Nature only for that of the being which she tends. Every selected character is fully exercised by her; and the being is placed under well-suited conditions of life. Man keeps the natives of many climates in the same country; he seldom exercises each selected character in some peculiar and fitting manner; he feeds a long and a short beaked pigeon on the same food; he does not exercise a long-backed or long-legged quadruped in any peculiar manner; he exposes sheep with long and short wool to the same climate. He does not allow the most vigorous males to struggle for the females. He does not rigidly destroy all inferior animals, but protects during each varying season, as far as lies in his power, all his productions. He often begins his selection by some half-monstrous form; or at least by some modification prominent enough to catch his eye, or to be plainly useful to him. Under nature, the slightest difference of structure or constitution may well turn the nicely-balanced scale in the struggle for life, and so be preserved. How fleeting are the wishes and efforts of man! how short his time! and consequently how poor will his products be, compared with those accumulated by nature during whole geological periods. Can we wonder, then, that nature's productions should be far "truer" in character than man's productions; that they should be infinitely better adapted to the most complex conditions of life, and should plainly bear the stamp of far higher workmanship?

It may be said that natural selection is daily and hourly scrutinising, throughout the world, every variation, even the slightest; rejecting that which is bad, preserving and adding up all that is good; silently and insensibly working, whenever and wherever opportunity offers, at the improvement of each organic being in relation to its organic and inorganic conditions of life. We see nothing of these slow changes in progress, until the hand of time has marked the long lapse of ages, and then so imperfect is our view into long past geological ages, that we only see that the forms of life are now different from what they formerly were.

Although natural selection can act only through and for the good of each being, yet characters and structures, which we are apt to consider as of very trifling importance, may thus be acted on. When we see leaf-eating insects green, and bark-feeders mottled-grey; the alpine ptarmigan white in winter, the red-grouse the colour of heather, and the black-grouse that of peaty earth, we must believe that these tints are of service to these birds and insects in preserving them from danger. Grouse, if not destroyed at some period of their lives, would increase in countless numbers; they are known to suffer largely from birds of prey; and hawks are guided by eyesight to their prey,—so much so, that on parts of the Continent persons are warned not to keep white pigeons, as being the most liable to destruction. Hence I can see no reason to doubt that natural selection might be most effective in giving the proper colour to each kind of grouse, and in keeping that colour, when once acquired, true and constant. Nor ought we to think that the occasional destruction of an animal of any particular colour would produce little effect: we should remember how essential it is in a flock of white sheep to destroy every lamb with the faintest trace of black. In plants the down on the fruit and the colour of the flesh are considered by botanists as characters of the most trifling importance: yet we hear from an excellent horticulturist, Downing, that in the United States smooth-skinned fruits suffer far more from a beetle, a curculio, than those with down; that purple plums suffer far more from a certain disease than yellow plums; whereas another disease attacks yellow-fleshed peaches far more than those with other coloured flesh. If, with all the aids of art, these slight differences make a great difference in cultivating the several varieties, assuredly, in a state of nature, where the trees would have to struggle with other trees and with a host of enemies, such differences would effectually settle which variety, whether a smooth or downy, a yellow or purple fleshed fruit, should succeed.

In looking at many small points of difference between species, which, as far as our ignorance permits us to judge, seem to be quite unimportant, we must not forget that climate, food, &c., probably produce some slight and direct effect. It is, however, far more necessary to bear in mind that there are many unknown laws of correlation of growth, which, when one part of the organisation is modified through variation, and the modifications are accumulated by natural selection for the good of the being, will cause other modifications, often of the most unexpected nature.

As we see that those variations which under domestication appear at any particular period of life, tend to reappear in the offspring at the same period;—for instance, in the seeds of the many varieties of our culinary and agricultural plants; in the caterpillar and cocoon stages of the varieties of the silkworm; in the eggs of poultry, and in the colour of the down of their chickens; in the horns of our sheep and cattle when nearly adult;—so in a state of nature, natural selection will be enabled to act on and modify organic beings at any age, by the accumulation of profitable variations at that age, and by their inheritance at a corresponding age. If it profit a plant to have its seeds more and more widely disseminated by the wind, I can see no greater difficulty in this being effected through natural selection, than in the cotton-planter increasing and improving by selection the down in the pods on his cotton-trees. Natural selection may modify and adapt the larva of an insect to a score of contingencies, wholly different from those which concern the mature insect. These modifications will no doubt affect, through the laws of correlation, the structure of the adult; and probably in the case of those insects which live only for a few hours, and which never feed, a large part of their structure is merely the correlated result of successive changes in the structure of their larvæ. So, conversely, modifications in the adult will probably often affect the structure of the larva; but in all cases natural selection will ensure that modifications consequent on other modifications at a different period of life, shall not be in the least degree injurious: for if they became so, they would cause the extinction of the species.

Natural selection will modify the structure of the

young in relation to the parent, and of the parent in relation to the young. In social animals it will adapt the structure of each individual for the benefit of the community; if each in consequence profits by the selected change. What natural selection cannot do, is to modify the structure of one species, without giving it any advantage, for the good of another species; and though statements to this effect may be found in works of natural history, I cannot find one case which will bear investigation. A structure used only once in an animal's whole life, if of high importance to it, might be modified to any extent by natural selection; for instance, the great jaws possessed by certain insects, and used exclusively for opening the cocoon—or the hard tip to the beak of nestling birds, used for breaking the egg. It has been asserted, that of the best short-beaked tumbler-pigeons more perish in the egg than are able to get out of it; so that fanciers assist in the act of hatching. Now, if nature had to make the beak of a full-grown pigeon very short for the bird's own advantage, the process of modification would be very slow, and there would be simultaneously the most rigorous selection of the young birds within the egg, which had the most powerful and hardest beaks, for all with weak beaks would inevitably perish: or, more delicate and more easily broken shells might be selected, the thickness of the shell being known to vary like every other structure.

Sexual Selection.—Inasmuch as peculiarities often appear under domestication in one sex and become hereditarily attached to that sex, the same fact probably occurs under nature, and if so, natural selection will be able to modify one sex in its functional relations to the other sex, or in relation to wholly different habits of life in the two sexes, as is sometimes the case with insects. And this leads me to say a few words on what I call Sexual Selection. This depends, not on a struggle for existence, but on a struggle between the males for possession of the females ; the result is not death to the unsuccessful competitor, but few or no offspring. Sexual selection is, therefore, less rigorous than natural selection. Generally, the most vigorous males, those which are best fitted for their places in nature, will leave most progeny. But in many cases, victory will depend not on general vigour, but on having special weapons, confined to the male sex. A hornless stag or spurless cock would have a poor chance of leaving offspring. Sexual selection by always allowing the victor to breed might surely give indomitable courage, length to the spur, and strength to the wing to strike in the spurred leg, as well as the brutal cock-fighter, who knows well that he can improve his breed by careful selection of the best cocks. How low in the scale of nature this law of battle descends, I know not; male alligators have been described as fighting, bellowing, and whirling round, like Indians in a war-dance, for the possession of the females; male salmons have been seen fighting all day long ; male stag-beetles often bear wounds from the huge mandibles of other males. The war is, perhaps, severest between the males of polygamous animals, and these seem oftenest provided with special weapons. The males of carnivorous animals are already well armed; though to them and to others, special means of defence may be given through means of sexual selection, as the mane to the lion, the shoulder-pad to the boar, and the hooked jaw to the male salmon; for the shield may be as important for victory, as the sword or spear.

Amongst birds, the contest is often of a more peaceful character. All those who have attended to the subject, believe that there is the severest rivalry between the males of many species to attract by singing the females. The rock-thrush of Guiana, birds of Paradise, and some others, congregate; and successive males display their gorgeous plumage and perform strange antics before the females, which standing by as spectators, at last choose the most attractive partner. Those who have closely attended to birds in confinement well know that they often take individual preferences and dislikes: thus Sir R. Heron has described how one pied peacock was eminently attractive to all his hen birds. It may appear childish to attribute any effect to such apparently weak means: I cannot here enter on the details necessary to support this view; but if man can in a short time give elegant carriage and beauty to his bantams, according to his standard of beauty, I can see no good reason to doubt that female birds, by selecting, during thousands of generations, the most melodious or beautiful males, according to their standard of beauty, might produce a marked effect. I strongly suspect that some well-known laws with respect to the plumage of male and female birds, in comparison with the plumage of the young, can be explained on the view of plumage having been chiefly modified by sexual selection, acting when the birds have come to the breeding age or during the breeding season; the modifications thus produced being inherited at corresponding ages or seasons, either by the males alone, or by the males and females; but I have not space here to enter on this subject.

Thus it is, as I believe, that when the males and females of any animal have the same general habits of life, but differ in structure, colour, or ornament, such differences have been mainly caused by sexual selection; that is, individual males have had, in successive generations, some slight advantage over other males, in their weapons, means of defence, or charms; and have transmitted these advantages to their male offspring. Yet, I would not wish to attribute all such sexual differences to this agency: for we see peculiarities arising and becoming attached to the male sex in our domestic animals (as the wattle in male carriers, horn-like protuberances in the cocks of certain fowls, &c.), which we cannot believe to be either useful to the males in battle, or attractive to the females. We see analogous cases under nature, for instance, the tuft of hair on the breast of the turkey-cock, which can hardly be either useful or ornamental to this bird;—indeed, had the tuft appeared under domestication, it would have been called a monstrosity.

[*Editor's Note:* Material has been omitted at this point.]

To sum up the circumstances favourable and unfavourable to natural selection, as far as the extreme intricacy of the subject permits. I conclude, looking to the future, that for terrestrial productions a large continental area, which will probably undergo many oscillations of level, and which consequently will exist for long periods in a broken condition, will be the most favourable for the production of many new forms of life, likely to endure long and to spread widely. For the area will first have existed as a continent, and the inhabitants, at this period numerous in individuals and kinds, will have been subjected to very severe competition. When converted by subsidence into large separate islands, there will still exist many individuals of the same species on each island: intercrossing on the confines of the range of each species will thus be checked: after physical changes of any kind, immigration will be prevented, so that new places in the polity of each island will have to be filled up by modifications of the old inhabitants; and time will be allowed for the varieties in each to become well modified and perfected. When, by renewed elevation, the islands shall be re-converted into a continental area, there will again be severe competition: the most favoured or improved varieties will be enabled to spread: there will be much extinction of the less improved forms, and the relative proportional numbers of the various inhabitants of the renewed continent will again be changed; and again there will be a fair field for natural selection to improve still further the inhabitants, and thus produce new species.

That natural selection will always act with extreme slowness, I fully admit. Its action depends on there being places in the polity of nature, which can be better occupied by some of the inhabitants of the country undergoing modification of some kind. The existence of such places will often depend on physical changes, which are generally very slow, and on the immigration of better adapted forms having been checked. But the action of natural selection will probably still oftener depend on some of the inhabitants becoming slowly modified; the mutual relations of many of the other inhabitants being thus disturbed. Nothing can be effected, unless favourable variations occur, and variation itself is apparently always a very slow process. The process will often be greatly retarded by free intercrossing. Many will exclaim that these several causes are amply sufficient wholly to stop the action of natural selection. I do not believe so. On the other hand, I do believe that natural selection will always act very slowly, often only at long intervals of time, and generally on only a very few of the inhabitants of the same region at the same time. I further believe, that this very slow, intermittent action of natural selection accords perfectly well with what geology tells us of the rate and manner at which the inhabitants of this world have changed.

Slow though the process of selection may be, if feeble man can do much by his powers of artificial selection, I can see no limit to the amount of change, to the beauty and infinite complexity of the coadaptations between all organic beings, one with another and with their physical conditions of life, which may be effected in the long course of time by nature's power of selection.

[*Editor's Note:* Material has been omitted at this point.]

17c

Reprinted from pages 200-206 of *On the Origin of Species by Means of Natural Selection, or the Preservation of Favoured Races in the Struggle for Life,* J. Murray, London, 1859

DIFFICULTIES ON THEORY

C. R. Darwin

Natural selection cannot possibly produce any modification in any one species exclusively for the good of another species; though throughout nature one species incessantly takes advantage of, and profits by, the structure of another. But natural selection can and does often produce structures for the direct injury of other species, as we see in the fang of the adder, and in the ovipositor of the ichneumon, by which its eggs are deposited in the living bodies of other insects. If it could be proved that any part of the structure of any one species had been formed for the exclusive good of another species, it would annihilate my theory, for such could not have been produced through natural selection. Although many statements may be found in works on natural history to this effect, I cannot find even one which seems to me of any weight. It is admitted that the rattlesnake has a poison-fang for its own defence and for the destruction of its prey; but some authors suppose that at the same time this snake is furnished with a rattle for its own injury, namely, to warn its prey to escape. I would almost as soon believe that the cat curls the end of its tail when preparing to spring, in order to warn the doomed mouse. But I have not space here to enter on this and other such cases.

Natural selection will never produce in a being anything injurious to itself, for natural selection acts solely by and for the good of each. No organ will be formed, as Paley has remarked, for the purpose of causing pain or for doing an injury to its possessor. If a fair balance be struck between the good and evil caused by each part, each will be found on the whole advantageous. After the lapse of time, under changing conditions of life, if any part comes to be injurious, it will be modified; or if it be not so, the being will become extinct, as myriads have become extinct.

Natural selection tends only to make each organic being as perfect as, or slightly more perfect than, the other inhabitants of the same country with which it has to struggle for existence. And we see that this is the degree of perfection attained under nature. The endemic productions of New Zealand, for instance, are perfect one compared with another; but they are now rapidly yielding before the advancing legions of plants and animals introduced from Europe. Natural selection will not produce absolute perfection, nor do we always meet, as far as we can judge, with this high standard under nature. The correction for the aberration of light is said, on high authority, not to be perfect even in that most perfect organ, the eye. If our reason leads us to admire with enthusiasm a multitude of inimitable contrivances in nature, this same reason tells us, though we may easily err on both sides, that some other contrivances are less perfect. Can we consider the sting of the wasp or of the bee as perfect, which, when used against many attacking animals, cannot be withdrawn, owing to the backward serratures, and so inevitably causes the death of the insect by tearing out its viscera?

If we look at the sting of the bee, as having originally existed in a remote progenitor as a boring and serrated instrument, like that in so many members of the same great order, and which has been modified but not perfected for its present purpose, with the poison originally adapted to cause galls subsequently intensified, we can perhaps understand how it is that the use of the sting should so often cause the insect's own death: for if on the whole the power of stinging be useful to the community, it will fulfil all the requirements of natural selection, though it may cause the death of some few members. If we admire the truly wonderful power of scent by which the males of many insects find their females, can we admire the production for this single purpose of thousands of drones, which are utterly useless to the community for any other end, and which are ultimately slaughtered by their industrious and sterile sisters? It may be difficult, but we ought to admire the savage instinctive hatred of the queen-bee, which urges her instantly to destroy the young queens her daughters as soon as born, or to perish herself in the combat; for undoubtedly this is for the good of the community; and maternal love or maternal hatred, though the latter fortunately is most rare, is all the same to the inexorable principle of natural selection. If we admire the several ingenious contrivances, by which the flowers of the orchis and of many other plants are fertilised through insect agency, can we consider as equally perfect the elaboration by our fir-trees of dense clouds of pollen, in order that a few granules may be wafted by a chance breeze on to the ovules?

Summary of Chapter.—We have in this chapter discussed some of the difficulties and objections which may be urged against my theory. Many of them are very grave; but I think that in the discussion light has been thrown on several facts, which on the theory of independent acts of creation are utterly obscure. We have seen that species at any one period are not indefinitely variable, and are not linked together by a multitude of intermediate gradations, partly because the process of natural selection will always be very slow, and will act, at any one time, only on a very few forms; and partly because the very process of natural selection almost implies the continual supplanting and extinction of preceding and intermediate gradations. Closely allied species, now living on a continuous area, must often have been formed when the area was not continuous, and when the conditions of life did not insensibly graduate away from one part to another. When two varieties are formed in two districts of a continuous area, an intermediate variety will often be formed, fitted for an intermediate zone; but from reasons assigned, the intermediate variety will usually exist in lesser numbers than the two forms which it connects; consequently the two latter, during the course of further modification, from existing in greater numbers, will have a great advantage over the less numerous intermediate variety, and will thus generally succeed in supplanting and exterminating it.

We have seen in this chapter how cautious we should be in concluding that the most different habits of life could not graduate into each other; that a bat, for instance, could not have been formed by natural selection from an animal which at first could only glide through the air.

We have seen that a species may under new conditions of life change its habits, or have diversified habits, with some habits very unlike those of its nearest congeners. Hence we can understand, bearing in mind that each organic being is trying to live wherever it can live, how it has arisen that there are upland geese with webbed feet, ground woodpeckers, diving thrushes, and petrels with the habits of auks.

Although the belief that an organ so perfect as the eye could have been formed by natural selection, is more than enough to stagger any one; yet in the case of any organ, if we know of a long series of gradations in complexity, each good for its possessor, then, under changing conditions of life, there is no logical impossibility in the acquirement of any conceivable degree of perfection through natural selection. In the cases in which we know of no intermediate or transitional states, we should be very cautious in concluding that none could have existed, for the homologies of many organs and their intermediate states show that wonderful metamorphoses in function are at least possible. For instance, a swim-bladder has apparently been converted into an air-breathing lung. The same organ having performed simultaneously very different functions, and then having been specialised for one function; and two very distinct organs having performed at the same time the same function, the one having been perfected whilst aided by the other, must often have largely facilitated transitions.

We are far too ignorant, in almost every case, to be enabled to assert that any part or organ is so unimportant for the welfare of a species, that modifications in its structure could not have been slowly accumulated by means of natural selection. But we may confidently believe that many modifications, wholly due to the laws of growth, and at first in no way advantageous to a species, have been subsequently taken advantage of by the still further modified descendants of this species. We may, also, believe that a part formerly of high importance has often been retained (as the tail of an aquatic animal by its terrestrial descendants), though it has become of such small importance that it could not, in its present state, have been acquired by natural selection,—a power which acts solely by the preservation of profitable variations in the struggle for life.

Natural selection will produce nothing in one species for the exclusive good or injury of another; though it may well produce parts, organs, and excretions highly useful or even indispensable, or highly injurious to another species, but in all cases at the same time useful to the owner. Natural selection in each well-stocked country, must act chiefly through the competition of the inhabitants one with another, and consequently will produce perfection, or strength in the battle for life, only according to the standard of that country. Hence the inhabitants of one country, generally the smaller one, will often yield, as we see they do yield, to the inhabitants of another and generally larger country. For in the larger country there will have existed more individuals, and more diversified forms, and the competition will have been severer, and thus the standard of perfection will have been rendered higher. Natural selection will not necessarily produce absolute perfection; nor, as far as we can judge by our limited faculties, can absolute perfection be everywhere found.

On the theory of natural selection we can clearly understand the full meaning of that old canon in natural history, " Natura non facit saltum." This canon, if we look only to the present inhabitants of the world, is not strictly correct, but if we include all those of past times, it must by my theory be strictly true.

It is generally acknowledged that all organic beings have been formed on two great laws—Unity of Type, and the Conditions of Existence. By unity of type is meant that fundamental agreement in structure, which we see in organic beings of the same class, and which is quite independent of their habits of life. On my theory, unity of type is explained by unity of descent. The expression of conditions of existence, so often insisted on by the illustrious Cuvier, is fully embraced by the principle of natural selection. For natural selection acts by either now adapting the varying parts of each being to its organic and inorganic conditions of life; or by having adapted them during long-past periods of time: the adaptations being aided in some cases by use and disuse, being slightly affected by the direct action of the external conditions of life, and being in all cases subjected to the several laws of growth. Hence, in fact, the law of the Conditions of Existence is the higher law; as it includes, through the inheritance of former adaptations, that of Unity of Type.

17D

Reprinted from pages 459–480 of *On the Origin of Species by Means of Natural Selection, or the Preservation of Favoured Races in the Struggle for Life,* J. Murray, London, 1859

RECAPITULATION AND CONCLUSION

C. R. Darwin

Recapitulation of the difficulties on the theory of Natural Selection — Recapitulation of the general and special circumstances in its favour — Causes of the general belief in the immutability of species — How far the theory of natural selection may be extended — Effects of its adoption on the study of Natural history — Concluding remarks.

As this whole volume is one long argument, it may be convenient to the reader to have the leading facts and inferences briefly recapitulated.

That many and grave objections may be advanced against the theory of descent with modification through natural selection, I do not deny. I have endeavoured to give to them their full force. Nothing at first can appear more difficult to believe than that the more complex organs and instincts should have been perfected, not by means superior to, though analogous with, human reason, but by the accumulation of innumerable slight variations, each good for the individual possessor. Nevertheless, this difficulty, though appearing to our imagination insuperably great, cannot be considered real if we admit the following propositions, namely,—that gradations in the perfection of any organ or instinct, which we may consider, either do now exist or could have existed, each good of its kind,—that all organs and instincts are, in ever so slight a degree, variable,—and, lastly, that there is a struggle for existence leading to the preservation of each profitable deviation of structure or instinct. The truth of these propositions cannot, I think, be disputed.

It is, no doubt, extremely difficult even to conjecture by what gradations many structures have been perfected, more especially amongst broken and failing groups of organic beings; but we see so many strange gradations in nature, as is proclaimed by the canon, "Natura non facit saltum," that we ought to be extremely cautious in saying that any organ or instinct, or any whole being, could not have arrived at its present state by many graduated steps. There are, it must be admitted, cases of special difficulty on the theory of natural selection; and one of the most curious of these is the existence of two or three defined castes of workers or sterile females in the same community of ants; but I have attempted to show how this difficulty can be mastered.

With respect to the almost universal sterility of species when first crossed, which forms so remarkable a contrast with the almost universal fertility of varieties when crossed, I must refer the reader to the recapitulation of the facts given at the end of the eighth chapter, which seem to me conclusively to show that this sterility is no more a special endowment than is the incapacity of two trees to be grafted together; but that it is incidental on constitutional differences in the reproductive systems of the intercrossed species. We see the truth of this conclusion in the vast difference in the result, when the same two species are crossed reciprocally; that is, when one species is first used as the father and then as the mother.

The fertility of varieties when intercrossed and of their mongrel offspring cannot be considered as universal; nor is their very general fertility surprising when we remember that it is not likely that either their constitutions or their reproductive systems should have been profoundly modified. Moreover, most of the varieties which have been experimentised on have been produced under domestication; and as domestication apparently tends to eliminate sterility, we ought not to expect it also to produce sterility.

The sterility of hybrids is a very different case from that of first crosses, for their reproductive organs are more or less functionally impotent; whereas in first crosses the organs on both sides are in a perfect condition. As we continually see that organisms of all kinds are rendered in some degree sterile from their constitutions having been disturbed by slightly different and new conditions of life, we need not feel surprise at hybrids being in some degree sterile, for their constitutions can hardly fail to have been disturbed from being compounded of two distinct organisations. This parallelism is supported by another parallel, but directly opposite, class of facts; namely, that the vigour and fertility of all organic beings are increased by slight changes in their conditions of life, and that the offspring of slightly modified forms or varieties acquire from being crossed increased vigour and fertility. So that, on the one hand, considerable changes in the conditions of life and crosses between greatly modified forms, lessen fertility; and on the other hand, lesser changes in the conditions of life and crosses between less modified forms, increase fertility.

Turning to geographical distribution, the difficulties encountered on the theory of descent with modification are grave enough. All the individuals of the same species, and all the species of the same genus, or even higher group, must have descended from common parents; and therefore, in however distant and isolated parts of the world they are now found, they must in the course of successive generations have passed from some one part to the others. We are often wholly unable even to conjecture how this could have been effected. Yet, as we have reason to believe that some species have retained the same specific form for very long periods, enormously long as measured by years, too much stress ought not to be laid on the occasional wide diffusion of the same species; for during very long periods of time there will always be a good chance for wide migration by many means. A broken or interrupted range may often be accounted for by the extinction of the species in the intermediate regions. It cannot be denied that we are as yet very ignorant of the full extent of the various climatal and geographical changes which have affected the earth during modern periods; and such changes will obviously have greatly facilitated migration. As an example, I have attempted to show how potent has been the influence of the Glacial period on the distribution both of the same and of representative species throughout the world. We are as yet profoundly ignorant of the many occasional means of transport. With respect to distinct species of the same genus inhabiting very distant and isolated regions, as the process of modification has necessarily been slow, all the means of migration will have been possible during a very long period; and consequently the difficulty of the wide diffusion of species of the same genus is in some degree lessened.

As on the theory of natural selection an interminable number of intermediate forms must have existed, linking together all the species in each group by gradations as fine as our present varieties, it may be asked, Why do we not see these linking forms all around us? Why are not all organic beings blended together in an inextricable chaos? With respect to existing forms, we should remember that we have no right to expect (excepting in rare cases) to discover *directly* connecting links between them, but only between each and some extinct and supplanted form. Even on a wide area, which has during a long period remained continuous, and of which the climate and other conditions of life change insensibly in going from a district occupied by one species into another district occupied by a closely allied species, we have no just right to expect often to find intermediate varieties in the intermediate zone. For we have reason to believe that only a few species are undergoing change at any one period; and all changes are slowly effected. I have also shown that the intermediate varieties which will at first probably exist in the intermediate zones, will be liable to be supplanted by the allied forms on either hand; and the latter, from existing in greater numbers, will generally be modified and improved at a quicker rate than the intermediate varieties, which exist in lesser numbers; so that the intermediate varieties will, in the long run, be supplanted and exterminated.

On this doctrine of the extermination of an infinitude of connecting links, between the living and extinct inhabitants of the world, and at each successive period between the extinct and still older species, why is not every geological formation charged with such links? Why does not every collection of fossil remains afford plain evidence of the gradation and mutation of the forms of life? We meet with no such evidence, and this is the most obvious and forcible of the many objections which may be urged against my theory. Why, again, do whole groups of allied species appear, though certainly they often falsely appear, to have come in suddenly on the several geological stages? Why do we not find great piles of strata beneath the Silurian system, stored with the remains of the progenitors of the Silurian groups of fossils? For certainly on my theory such strata must somewhere have been deposited at these ancient and utterly unknown epochs in the world's history.

I can answer these questions and grave objections only on the supposition that the geological record is far more imperfect than most geologists believe. It cannot be objected that there has not been time sufficient for any amount of organic change; for the lapse of time has been so great as to be utterly inappreciable by the human intellect. The number of specimens in all our museums is absolutely as nothing compared with the countless generations of countless species which certainly have existed. We should not be able to recognise a species as the parent of any one or more species if we were to examine them ever so closely, unless we likewise possessed many of the intermediate links between their past or parent and present states; and these many links we could hardly ever expect to discover, owing to the imperfection of the geological record. Numerous existing doubtful forms could be named which are probably varieties; but who will pretend that in future ages so many fossil links will be discovered, that naturalists will be able to decide, on the common view, whether or not these doubtful forms are varieties? As long as most of the links between any two species are unknown, if any one link or intermediate variety be discovered, it will simply be classed as another and distinct species. Only a small portion of the world has been geologically explored. Only organic beings of certain classes can be preserved in a fossil condition, at least in any great number. Widely ranging species vary most, and varieties are often at first local,—both causes rendering the discovery of intermediate links less likely. Local varieties will not spread into other and distant regions until they are considerably modified and improved; and when they do spread, if discovered in a geological formation, they will appear as if suddenly created there, and will be simply classed as new species. Most formations have been intermittent in their accumulation; and their duration, I am inclined to believe, has been shorter than the average duration of specific forms. Successive formations are separated from each other by enormous blank intervals of time; for fossiliferous formations, thick enough to resist future degradation, can be accumulated only where much sediment is deposited on the subsiding bed of the sea. During the alternate periods of elevation and of stationary level the record will be blank. During these latter periods there will probably be more variability in the forms of life; during periods of subsidence, more extinction.

With respect to the absence of fossiliferous formations beneath the lowest Silurian strata, I can only recur to the hypothesis given in the ninth chapter. That the geological record is imperfect all will admit; but that it is imperfect to the degree which I require, few will be inclined to admit. If we look to long enough intervals of time, geology plainly declares that all species have changed; and they have changed in the manner which my theory requires, for they have changed slowly and in a graduated manner. We clearly see this in the fossil remains from consecutive formations invariably being much more closely related to each other, than are the fossils from formations distant from each other in time.

Such is the sum of the several chief objections and difficulties which may justly be urged against my theory; and I have now briefly recapitulated the answers and explanations which can be given to them. I have felt these difficulties far too heavily during many years to doubt their weight. But it deserves especial notice that the more important objections relate to questions on which we are confessedly ignorant; nor do we know how ignorant we are. We do not know all the possible transitional gradations between the simplest and the most perfect organs; it cannot be pretended that we know all the varied means of Distribution during the long lapse of years, or that we know how imperfect the Geological Record is. Grave as these several difficulties are, in my judgment they do not overthrow the theory of descent with modification.

Now let us turn to the other side of the argument. Under domestication we see much variability. This seems to be mainly due to the reproductive system being eminently susceptible to changes in the conditions of life; so that this system, when not rendered impotent, fails to reproduce offspring exactly like the parent-form. Variability is governed by many complex laws,—by correlation of growth, by use and disuse, and by the direct action of the physical conditions of life. There is much difficulty in ascertaining how much modification our domestic productions have undergone; but we may safely infer that the amount has been large, and that modifications can be inherited for long periods. As long as the conditions of life remain the same, we have reason to believe that a modification, which has already been inherited for many generations, may continue to be inherited for an almost infinite number of generations. On the other hand we have evidence that variability, when it has once come into play, does not wholly cease; for new varieties are still occasionally produced by our most anciently domesticated productions.

Man does not actually produce variability; he only

unintentionally exposes organic beings to new conditions of life, and then nature acts on the organisation, and causes variability. But man can and does select the variations given to him by nature, and thus accumulate them in any desired manner. He thus adapts animals and plants for his own benefit or pleasure. He may do this methodically, or he may do it unconsciously by preserving the individuals most useful to him at the time, without any thought of altering the breed. It is certain that he can largely influence the character of a breed by selecting, in each successive generation, individual differences so slight as to be quite inappreciable by an uneducated eye. This process of selection has been the great agency in the production of the most distinct and useful domestic breeds. That many of the breeds produced by man have to a large extent the character of natural species, is shown by the inextricable doubts whether very many of them are varieties or aboriginal species.

There is no obvious reason why the principles which have acted so efficiently under domestication should not have acted under nature. In the preservation of favoured individuals and races, during the constantly-recurrent Struggle for Existence, we see the most powerful and ever-acting means of selection. The struggle for existence inevitably follows from the high geometrical ratio of increase which is common to all organic beings. This high rate of increase is proved by calculation, by the effects of a succession of peculiar seasons, and by the results of naturalisation, as explained in the third chapter. More individuals are born than can possibly survive. A grain in the balance will determine which individual shall live and which shall die,—which variety or species shall increase in number, and which shall decrease, or finally become extinct. As the individuals of the same species come in all respects into the closest competition with each other, the struggle will generally be most severe between them; it will be almost equally severe between the varieties of the same species, and next in severity between the species of the same genus. But the struggle will often be very severe between beings most remote in the scale of nature. The slightest advantage in one being, at any age or during any season, over those with which it comes into competition, or better adaptation in however slight a degree to the surrounding physical conditions, will turn the balance.

With animals having separated sexes there will in most cases be a struggle between the males for possession of the females. The most vigorous individuals, or those which have most successfully struggled with their conditions of life, will generally leave most progeny. But success will often depend on having special weapons or means of defence, or on the charms of the males; and the slightest advantage will lead to victory.

As geology plainly proclaims that each land has undergone great physical changes, we might have expected that organic beings would have varied under nature, in the same way as they generally have varied under the changed conditions of domestication. And if there be any variability under nature, it would be an unaccountable fact if natural selection had not come into play. It has often been asserted, but the assertion is quite incapable of proof, that the amount of variation under nature is a strictly limited quantity. Man, though acting on external characters alone and often capriciously, can produce within a short period a great result by adding up mere individual differences in his domestic productions; and every one admits that there are at least individual differences in species under nature. But, besides such differences, all naturalists have admitted the existence of varieties, which they think sufficiently distinct to be worthy of record in systematic works. No one can draw any clear distinction between individual differences and slight varieties; or between more plainly marked varieties and sub-species, and species. Let it be observed how naturalists differ in the rank which they assign to the many representative forms in Europe and North America.

If then we have under nature variability and a powerful agent always ready to act and select, why should we doubt that variations in any way useful to beings, under their excessively complex relations of life, would be preserved, accumulated, and inherited? Why, if man can by patience select variations most useful to himself, should nature fail in selecting variations useful, under changing conditions of life, to her living products? What limit can be put to this power, acting during long ages and rigidly scrutinising the whole constitution, structure, and habits of each creature,—favouring the good and rejecting the bad? I can see no limit to this power, in slowly and beautifully adapting each form to the most complex relations of life. The theory of natural selection, even if we looked no further than this, seems to me to be in itself probable. I have already recapitulated, as fairly as I could, the opposed difficulties and objections: now let us turn to the special facts and arguments in favour of the theory.

On the view that species are only strongly marked and permanent varieties, and that each species first existed as a variety, we can see why it is that no line of demarcation can be drawn between species, commonly supposed to have been produced by special acts of creation, and varieties which are acknowledged to have been produced by secondary laws. On this same view we can understand how it is that in each region where many species of a genus have been produced, and where they now flourish, these same species should present many varieties; for where the manufactory of species has been active, we might expect, as a general rule, to find it still in action; and this is the case if varieties be incipient species. Moreover, the species of the larger genera, which afford the greater number of varieties or incipient species, retain to a certain degree the character of varieties; for they differ from each other by a less amount of difference than do the species of smaller genera. The closely allied species also of the larger genera apparently have restricted ranges, and they are clustered in little groups round other species—in which respects they resemble varieties. These are strange relations on the view of each species having been independently created, but are intelligible if all species first existed as varieties.

As each species tends by its geometrical ratio of reproduction to increase inordinately in number; and as the modified descendants of each species will be enabled to increase by so much the more as they become more diversified in habits and structure, so as to be enabled to seize on many and widely different places in the economy of nature, there will be a constant tendency in natural selection to preserve the most divergent offspring of any one species. Hence during a long-continued course of modification, the slight differences, characteristic of varieties of the same species, tend to be augmented into the greater differences characteristic of species of the same genus. New and improved varieties will inevitably supplant and exterminate the older, less improved and intermediate varieties; and thus species are rendered to a large extent defined and distinct objects. Dominant species belonging to the larger groups tend to give birth to new and dominant forms; so that each large group tends to become still larger, and at the same time more divergent in character. But as all groups cannot thus succeed in increasing in size, for the world would not hold them, the more dominant groups beat the less dominant. This tendency in the large groups to go on increasing in size and diverging in character, together with the almost inevitable contingency of much extinction, explains the arrangement of all the forms of life, in groups subordinate to groups, all within a few great classes, which we now see everywhere around us, and which has prevailed throughout all time. This grand fact of the grouping of all organic beings seems to me utterly inexplicable on the theory of creation.

As natural selection acts solely by accumulating slight, successive, favourable variations, it can produce no great or sudden modification; it can act only by very short and slow steps. Hence the canon of "Natura non facit saltum," which every fresh addition to our knowledge tends to make more strictly correct, is on this theory simply intelligible. We can plainly see why nature is prodigal in variety, though niggard in innovation. But why this should be a law of nature if each species has been independently created, no man can explain.

Many other facts are, as it seems to me, explicable on this theory. How strange it is that a bird, under the form of woodpecker, should have been created to prey on insects on the ground; that upland geese, which never or rarely swim, should have been created with webbed feet; that a thrush should have been created to dive and feed on sub-aquatic insects; and that a petrel should have been created with habits and structure fitting it for the life of an auk or grebe! and so on in endless other cases. But on the view of each species constantly trying to increase in number, with natural selection always ready to adapt the slowly varying descendants of each to any unoccupied or ill-occupied place in nature, these facts cease to be strange, or perhaps might even have been anticipated.

As natural selection acts by competition, it adapts the inhabitants of each country only in relation to the degree of perfection of their associates; so that we need feel no surprise at the inhabitants of any one country, although on the ordinary view supposed to have been specially created and adapted for that country, being beaten and supplanted by the naturalised productions from another land. Nor ought we to marvel if all the contrivances in nature be not, as far as we can judge, absolutely perfect; and if some of them be abhorrent to our ideas of fitness. We need not marvel at the sting of the bee causing the bee's own death; at drones being produced in such vast numbers for one single act, and being then slaughtered by their sterile sisters; at the astonishing waste of pollen by our fir-trees; at the instinctive hatred of the queen bee for her own fertile daughters; at ichneumonidæ feeding within the live bodies of caterpillars; and at other such cases. The wonder indeed is, on the theory of natural selection, that more cases of the want of absolute perfection have not been observed.

The complex and little known laws governing variation are the same, as far as we can see, with the laws which have governed the production of so-called specific forms. In both cases physical conditions seem to have produced but little direct effect; yet when varieties enter any zone, they occasionally assume some of the characters of the species proper to that zone. In both varieties and species, use and disuse seem to have produced some effect; for it is difficult to resist this conclusion when we look, for instance, at the logger-headed duck, which has wings incapable of flight, in nearly the same condition as in the domestic duck; or when we look at the burrowing tucutucu, which is occasionally blind, and then at certain moles, which are habitually blind and have their eyes covered with skin; or when we look at the blind animals inhabiting the dark caves of America and Europe. In both varieties and species correlation of growth seems to have played a most important part, so that when one part has been modified other parts are necessarily modified. In both varieties and species reversions to long-lost characters occur. How inexplicable on the theory of creation is the occasional appearance of stripes on the shoulder and legs of the several species of the horse-genus and in their hybrids! How simply is this fact explained if we believe that these species have descended from a striped progenitor, in the same manner as the several domestic breeds of pigeon have descended from the blue and barred rock-pigeon!

On the ordinary view of each species having been independently created, why should the specific characters, or those by which the species of the same genus differ from each other, be more variable than the generic characters in which they all agree? Why, for instance, should the colour of a flower be more likely to vary in any one species of a genus, if the other species, supposed to have been created independently, have differently coloured flowers, than if all the species of the genus have the same coloured flowers? If species are only well-marked varieties, of which the characters have become in a high degree permanent, we can understand this fact; for they have already varied since they branched off from a common progenitor in certain characters, by which they have come to be specifically distinct from each other;

and therefore these same characters would be more likely still to be variable than the generic characters which have been inherited without change for an enormous period. It is inexplicable on the theory of creation why a part developed in a very unusual manner in any one species of a genus, and therefore, as we may naturally infer, of great importance to the species, should be eminently liable to variation; but, on my view, this part has undergone, since the several species branched off from a common progenitor, an unusual amount of variability and modification, and therefore we might expect this part generally to be still variable. But a part may be developed in the most unusual manner, like the wing of a bat, and yet not be more variable than any other structure, if the part be common to many subordinate forms, that is, if it has been inherited for a very long period; for in this case it will have been rendered constant by long-continued natural selection.

Glancing at instincts, marvellous as some are, they offer no greater difficulty than does corporeal structure on the theory of the natural selection of successive, slight, but profitable modifications. We can thus understand why nature moves by graduated steps in endowing different animals of the same class with their several instincts. I have attempted to show how much light the principle of gradation throws on the admirable architectural powers of the hive-bee. Habit no doubt sometimes comes into play in modifying instincts; but it certainly is not indispensable, as we see, in the case of neuter insects, which leave no progeny to inherit the effects of long-continued habit. On the view of all the species of the same genus having descended from a common parent, and having inherited much in common, we can understand how it is that allied species, when placed under considerably different conditions of life, yet should follow nearly the same instincts; why the thrush of South America, for instance, lines her nest with mud like our British species. On the view of instincts having been slowly acquired through natural selection we need not marvel at some instincts being apparently not perfect and liable to mistakes, and at many instincts causing other animals to suffer.

If species be only well-marked and permanent varieties, we can at once see why their crossed offspring should follow the same complex laws in their degrees and kinds of resemblance to their parents,—in being absorbed into each other by successive crosses, and in other such points,—as do the crossed offspring of acknowledged varieties. On the other hand, these would be strange facts if species have been independently created, and varieties have been produced by secondary laws.

If we admit that the geological record is imperfect in an extreme degree, then such facts as the record gives, support the theory of descent with modification. New species have come on the stage slowly and at successive intervals; and the amount of change, after equal intervals of time, is widely different in different groups. The extinction of species and of whole groups of species, which has played so conspicuous a part in the history of the organic world, almost inevitably follows on the principle of natural selection; for old forms will be supplanted by new and improved forms. Neither single species nor groups of species reappear when the chain of ordinary generation has once been broken. The gradual diffusion of dominant forms, with the slow modification of their descendants, causes the forms of life, after long intervals of time, to appear as if they had changed simultaneously throughout the world. The fact of the fossil remains of each formation being in some degree intermediate in character between the fossils in the formations above and below, is simply explained by their intermediate position in the chain of descent. The grand fact that all extinct organic beings belong to the same system with recent beings, falling either into the same or into intermediate groups, follows from the living and the extinct being the offspring of common parents. As the groups which have descended from an ancient progenitor have generally diverged in character, the progenitor with its early descendants will often be intermediate in character in comparison with its later descendants; and thus we can see why the more ancient a fossil is, the oftener it stands in some degree intermediate between existing and allied groups. Recent forms are generally looked at as being, in some vague sense, higher than ancient and extinct forms; and they are in so far higher as the later and more improved forms have conquered the older and less improved organic beings in the struggle for life. Lastly, the law of the long endurance of allied forms on the same continent,—of marsupials in Australia, of edentata in America, and other such cases,—is intelligible, for within a confined country, the recent and the extinct will naturally be allied by descent.

Looking to geographical distribution, if we admit that there has been during the long course of ages much migration from one part of the world to another, owing to former climatal and geographical changes and to the many occasional and unknown means of dispersal, then we can understand, on the theory of descent with modification, most of the great leading facts in Distribution. We can see why there should be so striking a parallelism in the distribution of organic beings throughout space, and in their geological succession throughout time; for in both cases the beings have been connected by the bond of ordinary generation, and the means of modification have been the same. We see the full meaning of the wonderful fact, which must have struck every traveller, namely, that on the same continent, under the most diverse conditions, under heat and cold, on mountain and lowland, on deserts and marshes, most of the inhabitants within each great class are plainly related; for they will generally be descendants of the same progenitors and early colonists. On this same principle of former migration, combined in most cases with modification, we can understand, by the aid of the Glacial period, the identity of some few plants, and the close alliance of many others, on the most distant mountains, under the most different climates; and likewise the close alliance of some of the inhabitants of the sea in the northern and southern temperate zones, though separated by the whole intertropical ocean. Although two areas may present the same physical conditions of life, we need feel no surprise at their inhabitants being widely different, if they have been for a long period completely separated from each other; for as the relation of organism to organism is the most important of all relations, and as the two areas will have received colonists from some third source or from each other, at various periods and in different proportions, the course of modification in the two areas will inevitably be different.

On this view of migration, with subsequent modification, we can see why oceanic islands should be inhabited by few species, but of these, that many should be peculiar. We can clearly see why those animals which cannot cross wide spaces of ocean, as frogs and terrestrial mammals, should not inhabit oceanic islands; and why, on the other hand, new and peculiar species of bats, which can traverse the ocean, should so often be found on islands far distant from any continent. Such facts as the presence of peculiar species of bats, and the absence of all other mammals, on oceanic islands, are utterly inexplicable on the theory of independent acts of creation.

The existence of closely allied or representative species in any two areas, implies, on the theory of descent with modification, that the same parents formerly inhabited both areas; and we almost invariably find that wherever many closely allied species inhabit two areas, some identical species common to both still exist. Wherever many closely allied yet distinct species occur, many doubtful forms and varieties of the same species likewise occur. It is a rule of high generality that the inhabitants of each area are related to the inhabitants of the nearest source whence immigrants might have been derived. We see this in nearly all the plants and animals of the Galapagos archipelago, of Juan Fernandez, and of the other American islands being related in the most striking manner to the plants and animals of the neighbouring American mainland; and those of the Cape de Verde archipelago and other African islands to the African mainland. It must be admitted that these facts receive no explanation on the theory of creation.

The fact, as we have seen, that all past and present organic beings constitute one grand natural system, with group subordinate to group, and with extinct groups often falling in between recent groups, is intelligible on the theory of natural selection with its contingencies of extinction and divergence of character. On these same principles we see how it is, that the mutual affinities of the species and genera within each class are so complex and circuitous. We see why certain characters are far more serviceable than others for classification; — why adaptive characters, though of paramount importance to the being, are of hardly any importance in classification; why characters derived from rudimentary parts, though of no service to the being, are often of high classificatory value; and why embryological characters are the most valuable of all. The real affinities of all organic beings are due to inheritance or community of descent. The natural system is a genealogical arrangement, in which we have to discover the lines of descent by the most permanent characters, however slight their vital importance may be.

The framework of bones being the same in the hand of a man, wing of a bat, fin of the porpoise, and leg of the horse,—the same number of vertebræ forming the neck of the giraffe and of the elephant,—and innumerable other such facts, at once explain themselves on the theory of descent with slow and slight successive modifications. The similarity of pattern in the wing and leg of a bat, though used for such different purpose,—in the jaws and legs of a crab,—in the petals, stamens, and pistils of a flower, is likewise intelligible on the view of the gradual modification of parts or organs, which were alike in the early progenitor of each class. On the principle of successive variations not always supervening at an early age, and being inherited at a corresponding not early period of life, we can clearly see why the embryos of mammals, birds, reptiles, and fishes should be so closely alike, and should be so unlike the adult forms. We may cease marvelling at the embryo of an air-breathing mammal or bird having branchial slits and arteries running in loops, like those in a fish which has to breathe the air dissolved in water, by the aid of well-developed branchiæ.

Disuse, aided sometimes by natural selection, will often tend to reduce an organ, when it has become useless by changed habits or under changed conditions of life; and we can clearly understand on this view the meaning of rudimentary organs. But disuse and selection will generally act on each creature, when it has come to maturity and has to play its full part in the struggle for existence, and will thus have little power of acting on an organ during early life; hence the organ will not be much reduced or rendered rudimentary at this early age. The calf, for instance, has inherited teeth, which never cut through the gums of the upper jaw, from an early progenitor having well-developed teeth; and we may believe, that the teeth in the mature animal were reduced, during successive generations, by disuse or by the tongue and palate having been fitted by natural selection to browse without their aid; whereas in the calf, the teeth have been left untouched by selection or disuse, and on the principle of inheritance at corresponding ages have been inherited from a remote period to the present day. On the view of each organic being and each separate organ having been specially created, how utterly inexplicable it is that parts, like the teeth in the embryonic calf or like the shrivelled wings under the soldered wing-covers of some beetles, should thus so frequently bear the plain stamp of inutility! Nature may be said to have taken pains to reveal, by rudimentary organs and by homologous structures, her scheme of modification, which it seems that we wilfully will not understand.

I have now recapitulated the chief facts and considerations which have thoroughly convinced me that species have changed, and are still slowly changing by the preservation and accumulation of successive slight favourable variations.

[*Editor's Note:* Material has been omitted at this point.]

18

Reprinted from pages 443–449 of *The Principles of Biology*, vol. 1, D. Appleton and Company, New York, 1884

INDIRECT EQUILIBRATION

H. Spencer

§ 164. Besides those perturbations produced in the moving equilibrium of any organism by special disturbing forces, there are ever going on many other perturbations—some which are the still-reverberating effects of disturbing forces previously experienced by the individual, and others which are the still-reverberating effects of disturbing forces experienced by ancestral individuals; and the multiplied deviations of function so caused, imply multiplied deviations of structure. In § 155 there was re-illustrated the truth, set forth at length when treating of Adaptation (§ 69), that an organism in a state of moving equilibrium, cannot have extra function thrown on any organ, and extra growth produced in such organ, without there being entailed correlative changes throughout all other functions, and eventually throughout all other organs. And when treating of Variation (§ 90), we saw that individuals which have been made, by their different circumstances, to deviate functionally and structurally from the average type in different directions, will bequeath to their joint offspring, compound perturbations of function and compound deviations of structure, endlessly varied in their kinds and amounts. That is to say, besides the primary perturbations and deviations directly caused in organisms by altered actions in their environments, there are ever being indirectly caused, secondary and tertiary perturbations and deviations, which, when compounded with one another from generation to generation, work innumerable slight modifications in the moving equilibria and correlative structures throughout the species.

Now if the individuals of a species are thus necessarily made unlike, in countless ways and degrees—if the complicated sets of rhythms which we call their functions, though similar in their general characters, are dissimilar in their details—if in one individual the amount of action in a particular direction is greater than in any other individual, or if here a peculiar combination gives a resulting force which is not found elsewhere; then, among all the individuals, some will be less liable than others to have their equilibria overthrown by a particular incident force, previously unexperienced. Unless the change in the environment is of so violent a kind as to be universally fatal to the species, it must affect more or less differently the slightly different moving equilibria which the members of the species present. It cannot but happen that some will be more stable than others, when exposed to this new or altered factor. That is to say, it cannot but happen that those individuals whose functions are most out of equilibrium with the modified aggregate of external forces, will be those to die; and that those will survive whose functions happen to be most nearly in equilibrium with the modified aggregate of external forces.

But this survival of the fittest, implies multiplication of the fittest. Out of the fittest thus multiplied, there will, as before, be an overthrowing of the moving equilibrium wherever it presents the least opposing force to the new incident force. And by the continual destruction of the individuals that are the least capable of maintaining their equilibria in presence of this new incident force, there must eventually be arrived at an altered type completely in equilibrium with the altered conditions.

§ 165. This survival of the fittest, which I have here

sought to express in mechanical terms, is that which Mr Darwin has called "natural selection, or the preservation of favoured races in the struggle for life." That there is going on a process of this kind throughout the organic world, Mr Darwin's great work on the *Origin of Species* has shown to the satisfaction of nearly all naturalists. Indeed, when once enunciated, the truth of his hypothesis is so obvious as scarcely to need proof. Though evidence may be required to show that natural selection accounts for everything ascribed to it, yet no evidence is required to show that natural selection has always been going on, is going on now, and must ever continue to go on. Recognizing this as an *à priori* certainty, let us contemplate it under its two distinct aspects.

That organisms which live, thereby prove themselves fit to live, in so far as they have been tried; while organisms which die, thereby prove themselves in some respects unfitted for living; are facts no less manifest, than is the fact that this self-acting purification of a species, must tend ever to insure adaptation between it and its environment. This adaptation may be either so *maintained* or so *produced*. Doubtless many who have looked at Nature with philosophic eyes, have observed that death of the worst and multiplication of the best, must result in the maintenance of a constitution in harmony with surrounding circumstances. That the average vigour of any race would be diminished, did the diseased and feeble habitually survive and propagate; and that the destruction of such, through failure to fulfil some of the conditions to life, leaves behind those which are able to fulfil the conditions to life, and thus keeps up the average fitness to the conditions of life; are almost self-evident truths. But to recognize "natural selection" as a means of preserving an already-established balance between the powers of a species and the forces to which it is subject, is to recognize it only in its simplest and most general mode of action. It is the more special mode of action with which we are here concerned. This more special mode of action, Mr Darwin has been the first to perceive. To him we owe the discovery that natural selection is capable of *producing* fitness between organisms and their circumstances; and he, too, has the merit of appreciating the immensely-important consequences that follow from this. He has worked up an enormous mass of evidence into an elaborate demonstration, that this "preservation of favoured races in the struggle for life," is an ever-acting cause of divergence among organic forms. He has traced out the involved results of the process with marvellous subtlety. He has shown how hosts of otherwise inexplicable facts, are fully accounted for by it. In brief, he has proved that the cause he alleges is a true cause; that it is a cause which we see habitually in action; and that the results to be inferred from it, are in harmony with the phenomena which the Organic Creation presents, both as a whole and in its details. Let us glance at a few of the more important interpretations which the hypothesis furnishes.

A soil possessing some ingredient in unusual quantity, may supply to a plant an excess of the matter required for a certain class of its tissues; and may cause all the parts formed of such tissues to be abnormally developed. Suppose that among these are the hairs clothing its surfaces, including those which grow on its seeds. Thus furnished with somewhat longer fibres, its seeds, when shed, are carried a little further by the wind before they fall to the ground. The young plants growing up from them, being rather more widely dispersed than those produced by other individuals of the same species, will be less liable to smother one another; and a greater number may therefore reach maturity and fructify. Supposing the next generation subject to the same peculiarity of nutrition, some of the seeds borne by its members will not simply inherit this increased development of hairs, but will carry it further; and these, still more advantaged in the same way as before, will, on the average, have still more numerous chances of continuing the race. Thus, by the survival, generation after generation, of those possessing these longer hairs, and the inheritance of successive increments of growth in the hairs, there may result a seed deviating greatly from the original. Other individuals of the same species, subject to the different physical conditions of other localities, may develop somewhat thicker or harder coatings to their seeds: so rendering their seeds less digestible by the birds that devour them. Such thicker-coated seeds, by escaping undigested more frequently than thinner-coated ones, will have additional chances of growing up and leaving offspring; and this process, acting in a cumulative manner through successive years, will produce a seed diverging in another direction from the ancestral type. Again, elsewhere, some modification in the physiologic actions of the plant, may lead to an unusual secretion of an essential oil in the seeds; which rendering them unpalatable to creatures that would otherwise feed on them, may diminish the destruction of the seeds, so giving an advantage to the variety in its rate of multiplication; and this incidental peculiarity proving a preservative, will, as before, be gradually increased by natural selection, until it constitutes another divergence. Now in these and countless analogous cases, we see that plants may become better adapted, or re-adapted, to the aggregate of surrounding agencies, not through any *direct* action of such agencies upon them, but through their *indirect* action—through the destruction by them of the individuals which are least congruous with them, and the survival of those which are most congruous with them. All these slight variations of function and structure, arising among the members of a species, serve as so many experiments; the great majority of which fail, but a few of which succeed. Just as we see that each plant bears a multitude of seeds, out of which some two or three happen to fulfil all the conditions required for reaching maturity, and continuing the race; so we see that each species is perpetually producing numerous slightly-modified forms, deviating in all directions from the average, out of which most fit the surrounding conditions no better than their parents, or not so well, but some few of which fit the conditions better; and doing so, are enabled the better to preserve themselves, and to produce offspring similarly capable of preserving themselves. Among animals the like process results in the like development of various structures which cannot have been affected by the performance of functions—their functions being purely passive. The thick shell of a mollusk, is inexplicable as a result of direct reactions of the organism against the external actions to which it is exposed; but it is quite explicable as a result of the survival, generation after generation, of individuals whose thicker coverings protected them against enemies. Similarly with such a dermal structure as that of the tortoise. Though we have evidence that the skin where it is continually exposed to pressure and friction may thicken, and so re-establish the equilibrium, by opposing a greater inner force to a greater outer force; yet we have no evidence that a coat of armour like that of the tortoise can be so produced. Nor, indeed, are the conditions under which only its production in such a manner could be accounted for, fulfilled; since the surface of the tortoise is not exposed to greater pressure and friction than the surfaces of other creatures. This massive carapace, and the strangely-adapted osseous frame-work which supports it, are unaccountable as results of evolution, unless through the process of natural selection. Thus, too, is it with the production of colours in birds and in insects; the formation of odoriferous glands in mammals; the growth of such excrescences as those of the camel. Thus, in short, is it with all those organs of animals, which do not play active parts in the compound rhythms of their functions.

Besides giving us explanations of structural characters that are otherwise unaccountable, Mr Darwin shows how natural selection explains peculiar relations between individuals in certain species. Such facts as the dimorphism of the primrose and other flowers, he proves to be quite in harmony with his hypothesis, though stumbling-blocks to all other hypotheses. While the production of neuters among bees and ants, is inexplicable as a result of direct adaptation, natural selection affords a feasible solution of it. The various differences that accompany difference of sex, sometimes slight, sometimes very great, are similarly accounted for. As before suggested (§ 79), natural selection appears capable of producing and maintaining the right proportion of the sexes in each species; and it requires but to contemplate the bearings of the argument, to see that the formation of different sexes may itself have been determined in the same way.

To convey here an adequate idea of Mr Darwin's doctrine, in the immense range of its applications, is of course impossible. The few illustrations just given, serving but dimly to indicate the many classes of phenomena interpreted by it, are set down simply to remind the reader what Mr Darwin's hypothesis is, and what are the else insoluble problems which it solves for us.

§ 166. But now, though it seems to me that we are thus supplied with a key to phenomena which are multitudinous and varied beyond all conception; it also seems to me that there is a moiety of the phenomena which this key will not unlock. Mr Darwin himself recognizes use and disuse of parts, as causes of modifications in organisms; and does this, indeed, to a greater extent than do some who accept his general conclusion. But I conceive that he does not recognize them to a sufficient extent. While he conclusively shows that the inheritance of changes of structure, caused by changes of function, is utterly insufficient to explain a great mass—probably the greater mass—of morphological phenomena: I think he leaves unconsidered a mass of morphological phenomena that are explicable as results of functionally-acquired modifications, transmitted and increased, and which are not explicable as results of natural selection.

[*Editor's Note:* Material has been omitted at this point.]

19

Reprinted from pages 267–271 of *More Letters of Charles Darwin: A Record of His Work in a Series of Hitherto Unpublished Letters,* vol. 1, F. Darwin and A. C. Seward, eds., J. Murray, London, 1903

CORRESPONDENCE ON THE USE OF HERBERT SPENCER'S "SURVIVAL OF THE FITTEST" METAPHOR TO DESCRIBE NATURAL SELECTION (JULY 2 AND 5, 1866)

A. R. Wallace and C. R. Darwin

A. R. Wallace to C. Darwin. Letter 190

Hurstpierpoint, Sussex, July 2nd, 1866.

I have been so repeatedly struck by the utter inability of numbers of intelligent persons to see clearly, or at all, the self-acting and necessary effects of Natural Selection, that I am led to conclude that the term itself, and your mode of illustrating it, however clear and beautiful to many of us, are yet not the best adapted to impress it on the general naturalist public. The two last cases of the misunderstanding are: (1) the article on "Darwin and his Teachings" in the last *Quarterly Journal of Science*, which, though very well written and on the whole appreciative, yet concludes with a charge of something like blindness, in your not seeing that Natural Selection requires the constant watching of an intelligent "chooser," like man's selection to which you so often compare it; and (2) in Janet's recent work on the *Materialism of the Present Day*, reviewed in last Saturday's *Reader*, by an extract from which I see that he considers your weak point to be that you do not see that "thought and direction are essential to the action of Natural Selection." The same objection has been made a score of times by your chief opponents, and I have heard it as often stated myself in conversation. Now, I think this arises almost entirely from your choice of the term "Natural Selection" and so constantly comparing it in its effects to Man's Selection, and also your so frequently personifying nature as "selecting," as "preferring," as

Letter 190 "seeking only the good of the species," etc., etc. To the few this is as clear as daylight, and beautifully suggestive, but to many it is evidently a stumbling-block. I wish, therefore, to suggest to you the possibility of entirely avoiding this source of misconception in your great work (if not now too late), and also in any future editions of the *Origin*, and I think it may be done without difficulty and very effectually by adopting Spencer's term (which he generally uses in preference to Natural Selection)—viz., "survival of the fittest."

This term is the plain expression of the fact; Natural Selection is a metaphorical expression of it, and to a certain degree indirect and incorrect, since, even personifying Nature, she does not so much select special variations as exterminate the most unfavourable ones.

Combined with the enormous multiplying powers of all organisms, and the "struggle for existence" leading to the constant destruction of by far the largest proportion—facts which no one of your opponents, as far as I am aware, has denied or misunderstood—"the survival of the fittest" rather than of those who were less fit could not possibly be denied or misunderstood. Neither would it be possible to say that to ensure the "survival of the fittest" any intelligent chooser was necessary; whereas when you say Natural Selection acts so as to choose those that are fittest, it *is* misunderstood, and apparently always will be. Referring to your book, I find such expressions as "Man selects only for his own good; Nature only for that of the being which she tends." This, it seems, will always be misunderstood; but if you had said "Man selects only for his own good; Nature, by the inevitable 'survival of the fittest,' only for that of the being she tends," it would have been less liable to be so.

I find you use the term "Natural Selection" in two senses: (1) for the simple preservation of favourable and rejection of unfavourable variations, in which case it is equivalent to "survival of the fittest"; and (2) for the effect or change produced by this preservation, as when you say, "To sum up the circumstances favourable or unfavourable to Natural Selection," and again, "Isolation, also, is an important element in the process of Natural Selection." Here it is not merely "survival of the fittest," but change produced by survival of the fittest, that is meant. On looking over your fourth chapter, I find that these alterations of terms can be in most cases easily made, while in some cases the addition of "or survival of the fittest" after "Natural Selection" would be best; and in others, less likely to be misunderstood, the original term may stand alone. Letter 190

I could not venture to propose to any other person so great an alteration of terms, but you, I am sure, will give it an impartial consideration, and if you really think the change will produce a better understanding of your work, will not hesitate to adopt it.

It is evidently also necessary not to personify "Nature" too much—though I am very apt to do it myself—since people will not understand that all such phrases are metaphors. Natural Selection is, when understood, so necessary and self-evident a principle, that it is a pity it should be in any way obscured; and it therefore seems to me that the free use of "survival of the fittest," which is a compact and accurate definition of it, would tend much to its being more widely accepted, and prevent it being so much misrepresented and misunderstood.

There is another objection made by Janet which is also a very common one. It is that the chances are almost infinite against the particular kind of variation required being coincident with each change of external conditions, to enable an animal to become modified by Natural Selection in harmony with such changed conditions; especially when we consider that, to have produced the almost infinite modifications of organic beings, this coincidence must have taken place an almost infinite number of times.

Now, it seems to me that you have yourself led to this objection being made, by so often stating the case too strongly against yourself. For example, at the commencement of Chapter IV. you ask if it is "improbable that useful variations should sometimes occur in the course of thousands of generations"; and a little further on you say, "unless profitable variations do occur, Natural Selection can do nothing." Now, such expressions have given your opponents the advantage of assuming that favourable variations are rare accidents, or may even for long periods never occur at all, and thus Janet's argument would appear to many to have

Letter 190 great force. I think it would be better to do away with all such qualifying expressions, and constantly maintain (what I certainly believe to be the fact) that variations of every kind are always occurring in every part of every species, and therefore that favourable variations are always ready when wanted. You have, I am sure, abundant materials to prove this; and it is, I believe, the grand fact that renders modification and adaptation to conditions almost always possible. I would put the burthen of proof on my opponents to show that any one organ, structure, or faculty does not vary, even during one generation, among all the individuals of a species; and also to show any mode or way in which any such organ, etc., does not vary. I would ask them to give any reason for supposing that any organ, etc., is ever absolutely identical at any one time in all the individuals of a species, and if not then it is always varying, and there are always materials which, from the simple fact that "the fittest survive," will tend to the modification of the race into harmony with changed conditions.

I hope these remarks may be intelligible to you, and that you will be so kind as to let me know what you think of them.

I have not heard for some time how you are getting on. I hope you are still improving in health, and that you will now be able to get on with your great work, for which so many thousands are looking with interest.

Letter 191

To A. R. Wallace.[1]

Down, July 5th [1866].

I have been much interested by your letter, which is as clear as daylight. I fully agree with all that you say on the advantages of H. Spencer's excellent expression of "the survival of the fittest." This, however, had not occurred to me till reading your letter. It is, however, a great objection to this term that it cannot be used as a substantive governing a verb; and that this is a real objection I infer from H. Spencer continually using the words Natural Selection. I formerly thought, probably in an exaggerated degree, that it was a great advantage to bring into connection natural and artificial selection; this indeed led me to use a term in

[1] From *Life and Letters*, III., p. 45.

common, and I still think it some advantage. I wish I had received your letter two months ago, for I would have worked in "the survival," etc., often in the new edition of the *Origin*, which is now almost printed off, and of which I will of course send you a copy. I will use the term in my next book on domestic animals, etc., from which, by the way, I plainly see that you expect *much* too much. The term Natural Selection has now been so largely used abroad and at home that I doubt whether it could be given up, and with all its faults I should be sorry to see the attempt made. Whether it will be rejected must now depend "on the survival of the fittest." As in time the term must grow intelligible the objections to its use will grow weaker and weaker. I doubt whether the use of any term would have made the subject intelligible to some minds, clear as it is to others; for do we not see even to the present day Malthus on Population absurdly misunderstood? This reflection about Malthus has often comforted me when I have been vexed at this misstatement of my views. As for M. Janet, he is a metaphysician, and such gentlemen are so acute that I think they often misunderstand common folk. Your criticism on the double sense in which I have used Natural Selection is new to me and unanswerable; but my blunder has done no harm, for I do not believe that any one, excepting you, has ever observed it. Again, I agree that I have said too much about "favourable variations," but I am inclined to think that you put the opposite side too strongly: if every part of every being varied, I do not think we should see the same end or object gained by such wonderfully diversified means. Letter 191

I hope you are enjoying the country, and are in good health, and are working hard at your *Malay Archipelago* book, for I will always put this wish in every note I write to you, as some good people always put in a text. My health keeps much the same, or rather improves, and I am able to work some hours daily.

20

Reprinted from *Linn. Soc. London Trans.* **23**:495–515 (1862)

CONTRIBUTIONS TO AN INSECT FAUNA OF THE AMAZON VALLEY. LEPIDOPTERA: *HELICONIDAE**

H. W. Bates

"Die wissenschaftliche Untersuchung der Natur strebt in den Einzelheiten das Allgemeine zu erkennen, um endlich dem Grunde aller Dinge näher zu kommen. Für diese Art Untersuchungen, die immer das Ziel der Naturforschung sein sollte, bietet wohl keine Thierclasse so reichen Stoff als die Insecten."—*Karl Ernst von Baer*, Address on the Opening of the Russian Entomological Society, St. Petersburg, May 1860.

THE family *Heliconidæ* was established by Mr. E. Doubleday in 1847, in Doubleday and Hewitson's 'Genera of Diurnal Lepidoptera.' It was founded on a number of Butterflies, remarkable for the elongated shape of their wings, and peculiar (with the exception of one genus, *Hamadryas*, which the author placed provisionally in the family, *op. cit.* p. 98) to the intertropical and subtropical zones of America. Many of them had been described by the older authors under *Heliconia*, *Mechanitis*, and several other ill-defined genera. They had been previously (in 1836) united in a tribe, *Heliconides*, by Dr. Boisduval in his 'Spécies Général des Lépidoptères;' but this comprehended also the group *Acræidæ*, which Doubleday excluded from the family. Linnæus treated them as a section of the genus *Papilio*, under the name of *Heliconii*. The nearest allies of the *Heliconidæ* are the *Acræidæ* just mentioned and the *Danaidæ*: all are distinguished from the true *Nymphalidæ* by the discoidal cell of the hind wings being always closed by perfect tubular nervules. Mr. Doubleday, placing more reliance on the shape of the antennæ and the abdominal border of the hind wings than on the far more important character above named, was led to exclude the genus *Eueides* from the family: this rendered the definition of the two groups very difficult, if not impossible, *Eueides* having the wing-cells closed in the same way as the *Heliconidæ*. Excepting that I re-admit *Eueides*, and exclude *Hamadryas*, which does not enter into the series of the American *Heliconidæ*, the family will be treated of in the present memoir as defined in the work above quoted.

The position of the *Heliconidæ* in the order Lepidoptera may be understood when I state that in a natural system the group would stand at the head of the whole series of families of which the order is composed. At least, this should be its place according to the view now taken of the order by many systematists, who arrange the families of *Rhopalocera*, or Butterflies, according to their degree of dissimilarity to the *Heterocera*, or Moths—in other words, according as their structure shows a lower or a higher stage in an ascending scale of organization. For, as the lower families of Moths are allied to other orders of insects, the further a group recedes from them in structure, the higher is the grade of perfection of the Lepidopterous type which it exhibits. The families show their degree of affinity to Moths by many characters, the principal of which is the

* The materials on which this memoir is founded were collected by the author during eleven years' research on the banks of the Amazons.

structure of the anterior legs in the adult state of the insects. The *Heterocera* have always six perfect legs: most of the families of *Rhopalocera* have the anterior pair in a more or less rudimentary condition; and as the atrophy seems to have reached its furthest stage in the *Heliconidæ*, this group must be considered as occupying the highest rank in the order. Other characters accompany the one derived from the structure of the legs, which it is unnecessary here to enumerate. It will be seen from these remarks that the order Lepidoptera is one of those groups in the Animal Kingdom which show, beyond the many collateral branches of development that always exist, a clear linear advancement of organization.

The *Heliconidæ*, *Danaidæ*, and *Acræidæ* are related to each other in a different way from that which appears in the received classifications. A few remarks on their mutual affinities are necessary, in order to exhibit the true relations of the *Heliconidæ* to the allied groups of the Old World tropics. It has escaped the notice of all authors, that the *Heliconidæ* are composed of two groups, which differ very considerably in important points of structure; in fact, the majority of the genera of which the family is composed ought to be withdrawn from it, and placed with the *Danaidæ*. The very great superficial resemblance between the two sets of genera has led to their being united by all authors, and prevented inquiry into their real relationship. To avoid innovation, I will retain the family as it stands, and call the group which is allied to the *Danaidæ*, DANAOID HELICONIDÆ, and the other, which approximates somewhat to the *Acræidæ*, ACRÆOID HELICONIDÆ. The Acræoid group comprehends the genera *Heliconius* and *Eueides*; the Danaoid, the whole of the remaining *Heliconidæ*. The following are the distinguishing characters of the two groups:—

Acræoid Heliconidæ. The hind wing-cell is very small, and the nervures are so arranged that the upper and lower radials* (discoidal nervures of Doubleday) appear to be branch and sub-branch of the subcostal nervure, the discocellulars being short and continuous with them; the costal nervure is prolonged to the apex of the wing. The larvæ are similar to those of *Acræa* and *Argynnis* (*Nymphalidæ*), being beset with hispid spines†. The head is broad, the palpi thick.

Danaoid Heliconidæ. The hind wing-cell is very large, and irregular in shape; the two radials never appear as branches of the subcostal, but are very uncertain in position, owing to the very vacillating length and direction of the discocellulars: the costal nervure is short, and terminates on the costa, not reaching the apex of the wing. The larvæ (only one species is known) are smooth, like those of the *Danaidæ*, but are furnished with tubercles, instead of long fleshy threads. Head small, orbicular; palpi slender‡.

* I have adopted the terminology of Doubleday (Doubl., Hewits., and Westwood's Genera of Diurnal Lepidoptera) with regard to the neuration or veining of the wings, excepting that I call the "nervules" of the subcostal and median nervures "branches," and the "discoidal nervures" "radials," these alterations appearing necessary to prevent the verbal confusion of nervule with nervure, and discoidal with discocellular.

† The early states of these insects were not known to Doubleday. I reared, myself, *Heliconius Erato* (and *Doris*) and *Eueides Lybia*. We are acquainted, through other sources, with the larvæ of *H. Melpomene*, *H. Ricini*, and *Mechanitis Polymnia*.

‡ Since the above memoir was read, Dr. C. Felder of Vienna, in an article entitled, "Specimen Faunæ Lepidopterologicæ riparum Negro superioris in Brasilia septentrionali," 'Wiener Entomologische Monatschrift,' March 1862,

This view of the affinities of the family will make a great difference in the conception of the group as regards the affiliation of the forms. Instead of being a group isolated in its structure, and peculiar to the tropical parts of America, it results that the bulk of the genera have a very close relationship to the *Danaidæ*, which are found in all hot countries of both hemispheres: the *Acræoid Heliconidæ* alone are an isolated set of forms. The American productions, however, show a great superiority in structure and in the diversity of the forms over those of the Old World. In Africa only one genus occurs, namely, *Danais*. In the tropical parts of Asia four genera are found, besides *Danais*; these are *Ideopsis*, *Hestia*, *Euplœa*, and *Hamadryas*, which contain together 74 species. In the hot parts of America 16 genera (of *Danaoid Heliconidæ*) have been discovered, comprising 233 described species. Besides this greater diversity of generic and specific forms, the American productions show a much greater advance in organization than those of the Old World; in other words, they recede further from what may be considered as the common type, namely, *Danais*. This is clear from the great and progressive modification in the position of the radial nervures and discocellulars of the hind wings, and the advanced stage of atrophy of the male fore legs reached by most of the genera. In all the Asiatic genera the fore legs are in the same condition as in *Danais* and the *Nymphalidæ*. The following Table will show the relative value of the productions of the two hemispheres in a clearer manner. I have placed the genera in accordance with what seem to be their mutual affinities. The relative length of the lines between them is a rough expression of the degree of relationship. The collateral lines of connexion are also attempted to be expressed.

DANAIDÆ.

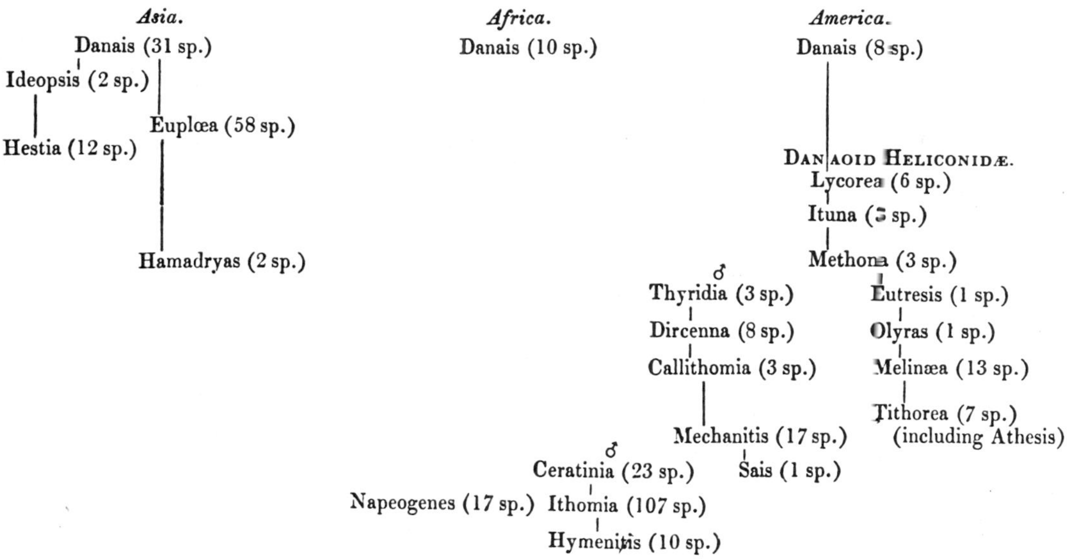

has ventured to withdraw the whole of the genera composing the group *Danaoid Heliconidæ*, placing them with the *Danaidæ*, and restricting the family *Heliconidæ* to the two genera *Heliconius* and *Eueides*. He has discovered an excellent and constant character for the *Danaidæ* (in which the *Danaoid Heliconidæ* participate), in addition to those already known, namely, the existence of a small nervule at the base of the fore-wing median nervure which ana-

There is a very wide dissimilarity in minor points and in general appearance between the Asiatic set of forms and the American: the only Old World genus which at all approaches the New World group is *Hamadryas*; but the shape, colours, and neuration of the wings show that it has no close affinity with them. The two sets of forms seem to agree, however, in habits, and apparently occupy the same sphere in the economy of nature in their respective countries. Mr. Wallace, who has had the good fortune to observe both in their native abodes, says, the habits of the South Asian *Euplœa* (the most numerous genus) are precisely those of the *Heliconidæ*. The Asiatic *Danaidæ* are mostly above the middle size, and include some of the largest Butterflies known; their American equivalents are in general below the middle size. Both are extremely prolific or abundant in individuals, and are amongst the most characteristic productions of their respective countries. Each set, also, are the objects of numerous mimetic resemblances on the part of other Lepidopterous insects of their own region belonging to different families,—the Asiatic mimickers being modelled after the Asiatic *Danaidæ*, and the American after the American members of the same family. The entire dissimilarity of the two sets of forms would seem to teach us that there can have been no land communication east and west between the tropical parts of Asia and America since they first came into existence, and therefore that the great continents must have remained separate in those quarters from a very remote epoch to allow for such an extensive independent development of forms. They are both strictly confined to the hottest parts of their respective hemispheres. In America they are not found beyond the northern tropic, nor much further south than 30° S. lat. They are not known to occur so far from the equator as either tropic in the Old World, but are limited to the south-eastern parts of Asia and the islands of the New Guinea group. The genus *Danais*, with which we have seen both groups are connected, ranges as far north as 41° in Europe, and 45° in North America. It is interesting thus to find that the only genus which is common to the three tropical regions is the sole one of the family that occurs in high latitudes. The only means of communication between the intertropical lands of America and Asia seems to have been a circuitous route by the north (or south); and the essentially tropical forms do not appear to have passed along it. The fact of the peculiar equatorial Asiatic *Danaidæ* not reaching Africa is explicable on the same grounds as their entire distinctness from the American ones, namely, the non-existence of an equatorial connexion of land of a nature suitable for their transit between the two continents since the remote date when the first forms of the group came into being.

The habits of the *Heliconidæ* have been described by various travellers,—Lacordaire having given a complete account of the Cayenne species, and Dyson and Gosse some interesting notes on those of Venezuela and Jamaica. The total number of species described is 284, namely, 233 belonging to the Danaoid, and 51 to the Acræoid group. They are peculiarly creatures of the forests, and, like the Platyrrhine Monkeys, the arboreal

stomoses with the median a short distance from its origin. In the systematic part of the present memoir I shall follow Dr. Felder in this altered classification. The two groups which composed the family *Heliconidæ* are, it must be repeated, completely and widely distinct. Yet the analogical resemblance between them is so great, that some species of the one might easily be confounded (if not closely examined) with species of the other.

Gallinacea (*Penelopidæ* and *Cracidæ*), and other groups of the same region, point to the gradual adaptation of the fauna, during an immense lapse of time, to a forest-clad country.

I found on the banks of the Amazons 94 species (besides many local varieties, considered by some authors as species) of the two groups (67 Danaoid and 27 Acræoid), representing all the genera of the family but three. They are most numerous in those parts of the country where the forests are most extensive and the climate most sultry and humid. I found the number of species to increase in travelling from east to west, from the Lower Amazons towards the eastern slopes of the Andes. They were rare in the somewhat drier tract of country which borders the Lower Amazons about the middle of its course. I found in this large district only 26 species, namely, 10 belonging to the Danaoid and 16 to the Acræoid group. Within an area of about the same dimensions, in the moist region of the Upper Amazons, I collected 64 species, of which 42 were Danaoid and 22 Acræoid *Heliconidæ*. I should judge, from the collections received in England from those parts, that the hot Andean valleys near Bogota, or in Ecuador, contain a still larger number of species than the plains of the Upper Amazons. In the dry forests which clothe a great part of the banks of the Tapajos I found exceedingly few: at one locality, where I collected four months, and which was rich in other families of Lepidoptera, I saw only one species of the Danaoid and four of the Acræoid group. According to Dyson, many species (*Ithomiæ*) of the lowlands in Venezuela have a vertical range of 2000 feet, and some genera (*Hymenitis* [*H. Dercetis*], *Olyras, Eutresis*), which do not inhabit the Amazon region, occur at an elevation of 8000 feet. The species are exceedingly abundant in individuals wherever they occur: they show every sign of flourishing existence, although of slow flight, feeble structure, unfurnished with apparent means of defence, and living in places which are incessantly haunted by swarms of insectivorous birds. The pathways in the forest near towns are quite enlivened by the multitudes which fly about amongst the lower trees, in their bright dresses of orange, blue, and yellow, and red and black.

The mode of flight of the members of the two groups is somewhat different. The *Heliconii* and *Eueides* move along in a sailing manner, often circling round for a considerable time, with their wings horizontally extended. The species of the Danaoid group, for the most part, keep near the ground, and have a very slow irregular flight, settling frequently. They are all of social or gregarious habits. Not only do individuals of the same species congregate in masses, but the set of closely allied species which people a district keep together in one or more compact flocks. I noticed in four districts rich in Danaoid *Heliconidæ*, where I collected, that about half the species of *Ithomia* flew together in one circumscribed area of the forest, and the other half in a second similar locality, the rest of the tolerably uniform wooded country, in each case, being nearly untenanted by them. The larger species (*Heliconii, Lycoreæ*) frequent flowers, probing the nectaries with their proboscides; but the smaller kinds (*Ithomiæ*), and the members of the Danaoid group generally, are very rarely found thus occupied: I noticed them sometimes imbibing drops of moisture from leaves and twigs. The fine showy *Heliconii* often assemble in small parties, or by twos and threes, apparently to sport together or perform a kind of dance. I believe the parties are composed chiefly of males. The sport begins generally between a single pair: they advance, retire, glide right and left in face of each other, wheel round to a considerable distance, again approach, and so on: a third joins in, then a fourth, or more. They never touch: when too many are congregated, a general flutter takes place, and they all fly off, to fall in again by pairs shortly afterwards. The species which I have seen most frequently employed in this way is the *Heliconius Rhea*, a glossy blue-black species, with bright yellow belts across its wings.

The larvæ of the two or three species whose transformations I observed feed together in clusters on the leaves of trees of moderate elevation, near the places where the adult insects are found.

The majority of the species have very limited ranges. I was surprised, when travelling on the Upper Amazons from east to west, to find the greater part of the species of *Ithomiæ* changed from one locality to another, not further removed than 100 to 200 miles. For instance, there were 11 of these *Ithomiæ* at a place called Fonte Boa, and 9 at St. Paulo, 180 miles distant; but only two of the total number (20) were found in both localities. This is remarkable when we consider that the whole of the country of the Upper Amazons is a nearly level plain, uniformly covered with forest, and offering no perceptible difference in soil or other physical conditions. Five only out of the 20 species have been met with in any other part of South America. The areas of distribution of most of the remaining 15 must be, in each case, a very limited tract of country. The species which inhabit other parts of Tropical America must have similarly contracted ranges, if we may judge from the collections received in England from different districts.

Now, many of these local species have the appearance of being geographical varieties; I could not help suspecting them to be such when I met with them in nature, the differences between the forms of one and those of another locality relating in many cases simply to the colours and colour-patterns of the wings. The marks of distinction, however, are in the majority so well defined, so ordinarily common to all the individuals concerned, and there is so generally an absence of connecting links, that they are held on all hands to be good and true species. Moreover, in those cases already mentioned, where a number of very closely allied species fly together, they keep themselves perfectly distinct; there are no hybrid forms (I am speaking of the *Ithomiæ* and allied genera), and on observing individuals *in copula*, I almost always * found the pair to be precisely the same in colours and markings. In the multiplicity, apparent distinctness, and restricted ranges of the species, this group much resembles the family of Humming-birds of the same regions.

I believe, nevertheless, that the suspicion of many of the species being nothing more than local modifications of other forms has proved to be well founded. Amongst the great number of perfectly distinct and well-marked species, a few occurred which showed great variability: these, I think, afford a key to the explanation of the origin of the rest. The details of variation will be given under the head of each species:

* The exception was in the case of *Mechanitis Polymnia*, which, as will be seen, on referring to the account of it in its place, is a polymorphic species, whose local varieties are in an imperfect state of segregation.

those which supply the most decisive results are *Mechanitis Polymnia, Ithomia Orolina* and *Illinissa, Ceratinia Ninonia*, and the *Lycoreæ*. The varieties of these present all the different grades between simple individual differences and well-marked local varieties or races, which latter cannot be distinguished from true species, when two or more of them are found coexisting in the same locality without intercrossing, as takes place in *Ithomia Illinissa* and its allies, and probably in *Mechanitis Nesæa* and *Lysimnia*. A striking case of the production of a local variety now spread over a wide area, and undistinguishable from a true species, is afforded in *Heliconius Thelxiope*, to the details of which, given in the systematic part of this memoir, I must refer the reader.

These species, when carefully studied, seem to me conclusively to show that many of the now distinct species of *Heliconidæ* have arisen from local varieties, segregated from the variations of preexisting widely disseminated species; for these distinct forms or species do not essentially differ from the undoubted varieties of the species cited. The genera show different degrees of susceptibility of change under altered local conditions. Thus, many species of *Heliconius* (*H. Rhea, Clytia, Ricini, Vesta, Thelxiope, Antiocha*, &c.) are unchanged over the whole of the wide country which includes the areas of several successive local races of many *Ithomiæ* and *Napeogenes* *.

The process of the creation of a new species I believe to be accelerated in the *Ithomiæ* and allied genera by the strong tendency of the insects, when pairing, to select none but their exact counterparts: this also enables a number of very closely allied ones to exist together, or the representative forms to live side by side on the confines of their areas, without amalgamating.

The course followed by Nature in the formation of these numerous local species, I think, is clearly exhibited in *Mechanitis Polymnia*, to the details of which, given in its place, I must beg the reader to refer. We see here the manufacture, as it were, in process. The species is widely disseminated and variable. The external conditions in certain localities are more favourable to one or more of the varieties there existing than to the others; those favoured ones, therefore, prevail over the others. We find, in this most instructive case, all the stages of the process, from the commencement of the formation of a local variety (var. *Egaënsis*) to the perfect segregation of one (var. *Lysimnia*, considered by all authors as a true species). In this species, most of the local varieties are connected with their parent form by individuals exhibiting all the shades of variation; and it is on this account only that we know them to be varieties. In the species allied to *Ithomia Flora*, the forms are in a complete state of segregation (with the exception of *I. Illinissa*, which throws light on the rest), and therefore they are considered as species; they are, in fact, perfectly good species, like all other forms considered as such in natural history. It is only by the study of variable species that we can obtain a clue to the explanation of the rest. But such species must be studied in nature, and with

* These are *Ithomia Flora*, an inhabitant of the whole Lower Amazon region (from the Atlantic to the Rio Negro), which is represented by *I. Hippodamia* in Cayenne, and *I. Onega, Illinissa, Gunilla, Priscilla, Ilerdina*, in different areas on the Upper Amazon; *Napeogenes Cyrianassa*, which becomes *N. adelphe* on the banks of the Cupari (Tapajos), and *N. Tunantina* on the north bank of the Upper Amazons; *N. Inachia*, which is changed to *N. sulphurina* at Bahia, and to *N. Ercilla, N. Corena* and *N. Pharo* in different areas on the Upper Amazon. Other species might be added in confirmation. Most of the species of *Heliconius* quoted are found unchanged over the collective areas of all these forms of *Ithomia* and *Napeogenes*.

strict reference to the *geographical relations* of their varieties. Many closet naturalists, who receive disconnectedly the different varieties in any group, treat them all as independent species: by such a proceeding, it is no wonder that they have faith in the absolute distinctness and immutability of species.

The sexes in the *Heliconidæ* very rarely differ in colours. Secondary sexual characters of another description occur, however, very generally in the Danaoid group. The males, in all the genera but two (*Lycorea* and *Ituna*) of this section, are furnished with a pencil or fringe of long hairs near the costal edge of the hind wings on the upper surface. It sometimes arises from the bottom of a shallow horny cup situated between the costal and subcostal nervures; the hairs are long, soft, and adpressed. I was unable to discover any use in this structure; it seemed not to be under the control of the insect. There is no movement in flight, or position in repose, peculiar to the male sex, which might require an instrument to hold the wings together—a function which the position of the hairs, in the place where the fore wing overlaps the hind wing, suggests to the mind. I believe the appendage must be considered as an outgrowth of the male organization, which is not in this case applied to any especial purpose: it may be taken to be of the same nature as the pencil of hairs on the breast of the male Turkey. Growths of one kind or other, on the surface of the wings, peculiar to the male sex, are frequent in Butterflies: in *Danais* the males have a small horny excrescence on the disk of the hind wings, which, considering the near relationship proved to exist between the two groups, I take to be homologically the same as the pencil of hairs in the *Danaoid Heliconidæ*. In the genus *Pavonia*, belonging to the family *Brassolidæ*, the males in some species have a fringe of hairs near the abdominal border; in others, a long pencil of the same on the disk; and, again, in others, instead of these appendages, a thickened plate on the inner margin of the hind wings.

The most interesting part of the natural history of the *Heliconidæ* is the mimetic analogies of which a great many of the species are the objects. Mimetic analogies, it is scarcely necessary to observe, are resemblances in external appearance, shape, and colours between members of widely distinct families: an idea of what is meant may be formed by supposing a Pigeon to exist with the general figure and plumage of a Hawk. Most modern authors who have written on the group have mentioned the striking instances of this kind of resemblances exhibited with reference to the *Heliconidæ*; but no attempt has been made to describe them fully, nor to explain them. I will give a short account of the leading facts, and then mention some circumstances which seem to throw light on their true nature and origin.

A large number of the species are accompanied in the districts they inhabit by other species which counterfeit them in the way described. The imitators belong to the following groups:—*Papilio*, *Pieris*, *Euterpe*, and *Leptalis* (fam. *Papilionidæ*), *Protogonius* (*Nymphalidæ*), *Ithomeis* (*Erycinidæ*), *Castnia* (*Castniadæ*), *Dioptis*, *Pericopis*, *Hyelosia*, and other genera (*Bombycidæ* Moths)*. I conclude that the *Heliconidæ* are the *objects imitated*, because they all have the same family facies, whilst the analogous species are dissimilar to their nearest allies—perverted, as it were, to produce the resemblance, from

* The accompanying Table, in which a number of the most striking of these are arranged in parallel columns, will give some idea of the extent to which this system of imitation prevails.

Specific Mimetic Analogies in the Lepidoptera of Tropical America, especially between the dominant family Heliconidæ and different other families.

Danaoid Heliconidæ.	Acræoid Heliconidæ.	Acræidæ.	Papilioninæ.	Pierinæ.	Erycinidæ.	Castniadæ.	Bombycidæ.
Lycorea Atergatis	...	...	Papilio Zagreus*	Euterpe Eurytele*	...	...	Pericopis angulata.
Methona Psidii (Ituna Phenarete*; I. Ilione*)	...	...	...	Leptalis Orise*	...	Castnia Linus*	P. *sp.*
Dircenna Epidero	...	...	...	...	...	...	Hyelosia Tiresia.
Tithorea Bonplandii	Heliconius Hecuba.*						
Melinæa Egina	H. Sylvana.						
M. Pardalis	H. Pardalinus*.						
M. Lucifer	H. Aurora.						
M. Messatis	H. Ismenius.						
M. (?) Hezia	H. Zuleika.						
M. Mnasias*—Ceratinia Ninonia.							
M. Mneme	H. Numata.						
Mechanitis Egaënsis	H. Eucoma, var.	...	...	L. Egaëna*.			
M. Nesæa—Napeogenes Xanthone*.	H. Ethra	...	...	L. Astyoche*.			
M. Lysimnia	H. Eucrate.						
Callithomia Alexirhoë* — Ceratinia Ninonia.							
Ithomia Flora (I. Cymo; Napeogenes Ithra*)	...	...	...	L. Theonoë*	Ithomeis satellites*	...	Dioptis Cyma.
I. Phono	...	...	...	...	...	...	D. diaphana.
I. Illinissa and allies) Napeogenes Ercilla*	...	...	...	L. Lysinoë, *var.**	I. aurantiaca*	...	D. Æliana.
I. Chrysodonia and allies) N. Corena*	...	...	...	L. Erythroë* *var.* of Theonoë.	I. mimica*	...	D. *sp.**
I. Ilerdina	...	...	...	L. Leuconoë* *var.* of Theonoë.	...	...	Dioptis Ilerdina*) Pericopis, *sp.**
Ceratina Fluonia	...	...	...	...	...	...	Dioptis, *sp.**
Ithomia Onega	...	...	...	L. Melanoë* *var.* of Theonoë.	I. Heliconina*	...	Dioptis Onega.
I. Virginia	...	...	...	L. Argochloë*. *var.* of Theonoë.			
I. Primula—Napeogenes Pharo.							
I. Euritæa—N. sulphurina.							
I. Eurimedia	...	...	...	L. Eumelia.			
I. Celemia—Napeogenes Tolosa.							
Ceratinia villula—N. Apulia	...	...	...	L. Siloë*.			
	Eueides Lampeto*	...	...	...	Stalachtis Calliope	...	Pericopis, *sp.**
	Heliconius Thelxiope	...	...	...	...	...	P. Isso*.
	H. Clytia	...	P. Pausanias*.				
	H. Vesta (Eueides Eanes*; E. Thales.)						
		Acræa Thalia.	...	L. Acreoides*	...	C. Acreoides*.	
			P. Zacynthus ♀	Euterpe Tereas.			
			P. Æneas and allies ♀	...	...	...	Pericopis turbida*.
			P. Bolivar ♀	...	...	C. *sp.**	
				Leptalis Lysinoë* *var.* of Theonoë	Ithomeis stalachtina*) Stalachtis Duvalii.		

* Those species marked * have been ascertained beyond all doubt to be very much fewer in individuals than the species which they mimic.

the normal facies of the genus or family to which they severally belong*. The resemblance is so close, that it is only after long practice that the true can be distinguished from the counterfeit, when on the wing in their native forests. I was never able to distinguish the *Leptalides* from the species they imitated, although they belong to a family totally different in structure and metamorphosis from the *Heliconidæ*, without examining them closely after capture. They fly in the same parts of the forest, and generally in company with the species they mimic.

I have already given an account of the local modifications to which the *Heliconidæ* are subject. It is a most curious circumstance, that corresponding races or species of counterfeiting groups accompany these local forms. In some cases I found proof that such species are modified from place to place to suit the peculiar forms of *Heliconidæ* there stationed. As this is an important point, and one which throws light on the origin of mimetic species, I must ask the reader's careful attention to the details, referring to the plates.

Plate LV. fig. 1 *a* (*Ithomia Flora*) and fig. 1 (*Leptalis Theonöe*) represent a Heliconide and its imitator, both of which inhabit the banks of the Cuparí, a river belonging to the Amazon system, in 55° W. long. Neither of these is found on the Upper Amazons (60° to 70° W. long.), where I made the remaining part of my observations on these insects. At Ega, on this upper river, in 65° W. long., two species of *Ithomia* occurred, which I consider to be local varieties or races of *I. Flora*, namely, *I. Onega* (Pl. LV. fig. 2 *a*) and *I. Illinissa* (Pl. LV. fig. 6 *a*). It is immaterial to the question in hand whether these be considered absolutely distinct species or races; the *Leptalis* which was found in their company was the form called *L. Lysinoë* (Pl. LV. fig. 3), with its admitted varieties (figs. 4, 5, 6, and 8). Only one of these varieties of *Leptalis* mimics an *Ithomia*; this is our fig. 6, which evidently counterfeits *Ithomia Illinissa* (fig. 6 *a*). The prevailing form of *Leptalis*, the *L. Lysinoë* (fig. 3), has no resemblance to any *Ithomia* of Ega, but is, when flying, a wonderful imitation of the *Stalachtis Duvalii* (Pl. LV. fig. 3 *a*), a common insect belonging to a genus (family *Erycinidæ*) equally flourishing and abundant in individuals with the members of the family *Heliconidæ*. I think there will be no doubt in the mind of any one that the Ega *Leptalides* are local varieties of the Cuparí *L. Theonoë* (fig. 1), when all the connecting links between them are studied in the figures given on our two plates. It is highly probable, therefore, that this species has been by some means modified with especial reference to the changed *Ithomiæ*, or other insects, of the locality. The varieties, figs. 4, 5, and 8, were excessively rare: they have the appearance of *sports*, and show how variable the species has been in this district.

The same takes place at St. Paulo, in 69° W. long. Here we find the *Ithomiæ* again changed. Neither the *I. Flora* of the Cuparí and Lower Amazons nor the *I. Illinissa* of Ega occurs; but the second Ega species, *I. Onega*, inhabits the district, and several other species not found in other places, amongst them *I. Ilerdina* (Pl. LVI. fig. 4 *a*), *I. Chrysodonia* (Pl. LVI. fig. 3 *a*), and *I. Virginia* (Pl. LVI. fig. 6 *a*). The prevailing species of

* This may be seen from the figures given of *Leptalis*,—fig. 5, Pl. LVI. being *L. Nehemia*, a species exhibiting the usual form of the family *Pieridæ*, to which the genus *Leptalis* belongs; whilst all the other *Leptalides* figured are mimetic species, totally unlike, as far as facies is concerned, this normal form.

Ithomia of the locality being thus changed, how stands it with the *Leptalides*? They are changed also, and again with close reference to the *Ithomiæ*. I found a number of different varieties, which I could not doubt were local forms of the same species as that found on the Cuparí and at Ega. Thus, there was one (Pl. LV. fig. 2) closely resembling *L. Theonoë* (fig. 1), but modified to produce a nearer imitation of the *Ithomia Onega* (Pl. LV. fig. 2 *a*), which I believe to be a local form of *I. Flora*. Another (Pl LVI. fig. 3) resembled *Ithomia Chrysodonia* (Pl. LVI. fig. 3 *a*); but the imitation is not fixed or exact in all the specimens taken, as may be seen by comparing figs. 1, 2, 3, of the same Plate. We here detect nature, as it were, *striving* after a correct imitation: the explanation of this will be attempted further on. A third form of *Leptalis* found at St. Paulo is the one figured Pl. LVI. fig. 6, which mimics the *Ithomia Virginia* (Pl. LVI. fig. 6 *a*). Besides these, a few varieties occurred which did not closely counterfeit any *Ithomia*; they were very much rarer than the others. I figure two of these (Pl. LV. figs. 7, 9), to show how they connect the other more strongly modified varieties with the Ega forms.

The *Ithomiæ* concerned in these imitations have the character of true species, being distinct and constant, with the exception of *I. Chrysodonia*, whose varieties are detailed under the head of the species, which is variable, and throws light on the origin of the rest. They are all excessively numerous in individuals, swarms of each kind being found in the localities they inhabit. The *Leptalides* are exceedingly rare; they cannot be more than as 1 to 1000 with regard to the *Ithomiæ*. It may be asked, how can we know they are all varieties (using the term as meaning forms descended from others) of one species? I must refer to the figures given, which, although they do not include all the connecting varieties that were collected, show how nearly all the forms are linked together. The most distinct amongst them are those figured Pl. LVI. figs. 4 and 6. The feature which distinguishes fig. 4 is the white colour of the disk of the hind wings, and the veins which traverse it. This character is shown to be due to variation, from the facts that *Ithomia Oncidia*, an undoubted variety of *I. Chrysodonia* (or *Orolina*), exhibits a commencement of this milky shade of the wings, and that many individuals of *I. Ilerdina* (Pl. LVI. fig. 4 *a*) display steps of modification in the colours of the veins. The variety figured Pl. LVI. fig. 6, appears distinct, from the single pale spot near the tips of the wings; an approximation to this is seen in the variety figured Pl. LV. fig. 9, which is an undoubted modification of *L. Lysinoë* (Pl. LV. fig. 3). The remarkable variety figured Pl. LV. fig. 4 has been described by the only author who has treated on these insects (Mr. Hewitson) as a variety of *L. Lysinoë*. In a polymorphic form, like this *Leptalis*, none of the varieties can be taken from the rest and denominated species, (using the term as meaning forms which cannot have descended from other closely allied ones), without exercising the art of species-making in the most arbitrary manner. For if we allow so great a latitude to variation as that from figs. 3 to 4, 5, 6, 7, 8, and 9, Pl. LV., how can we venture to say that natural modification, having gone so far, was incompetent to go further, so as to produce figs. 4 and 6, Pl. LVI., and that those forms must have arisen by some unknown agency? It is true, they have not arisen by simple variation, or *sports*, in one generation, but, as we shall presently see, by an external agency accumulating the modifications of many generations in two diverging directions. As the connecting links have not all been found, they may be called species: the word is of little importance. The habits of all are the same. When I had collected only two or three of the most distinct, I considered them separate species; but intermediate forms successively occurred, every capture tending to link the whole more closely together. The explanation that the whole are the result of hybridization from a few originally distinct species cannot at all apply in this case, because the distinct forms whose intercrossing would be required to produce the hybrids are confined to districts situated many hundred miles apart.

None of these *Leptalides* have been found in any other district or country than those inhabited by the *Ithomiæ* which they counterfeit. A species very closely allied to *L. Lysinoë*, var. *Argochloë* (Pl. LVI. fig. 6), has been received from Mexico (*L. Antherize*); but an *Ithomia*, of nearly the same colours (*I. Nero*) also inhabits Mexico. Many other species of *Leptalis*, of much larger size than the one here discussed, also mimic *Heliconidæ*, the objects of imitation not being *Ithomiæ*, but other genera of the family. Two of these are figured on Pl. LVI. *L. Orise* (Pl. LVI. fig. 8) is a remarkably exact counterfeit of *Methona Psidii* (fig. 8 *a*), the resemblance being carried to minutiæ, such as the colour of the antennæ and the spotting of the abdomen. *L. Amphione*, var. *Egaëna* (Pl. LVI. fig. 7), is very curious, as being a satellite of *Mechanitis Polymnia*, var. *Egaënsis* (fig. 7 *a*), both peculiar to the district of Ega,—the typical *L. Amphione* being found at Surinam, in company with the typical *M. Polymnia*, which it resembles—local varieties or sister species of *Leptalis Amphione* accompanying local varieties of *Mechanitis Polymnia* in other parts of tropical America.

Several species of *Dioptis*, a genus of Moths, and *Ithomeis*, a genus of *Erycinidæ*, also accompany these species or distinct local forms of *Ithomia*. A few of the Moths are figured on Pl. LV. figs. 10, 11, 12, 13. The imitations may not appear very exact from the figures; but when the insects are seen on the wing in their native woods, they deceive the most experienced eye.

A similar series of mimetic analogies occurs in the Old World, between the Asiatic and African *Danaidæ*, or representatives of the *Heliconidæ*, and species of other families of Butterflies and Moths. No instance is known in these families of a tropical species of one hemisphere counterfeiting a form belonging to the other. A most remarkable case of mimicry has been recorded by Mr. Trimen* in a *Papilio* of Southern Africa, *P. Cenea*, whose male wears to deception the livery of one species of *Danais*, namely, *D. Echeria*, whilst the female resembles a quite different one, *D. Chrysippus*,—both African. Mimetic analogies, however, are not confined to the Lepidoptera; most orders of insects supply them; but they are displayed only by certain families. Many instances are known where parasitic Bees and two-winged Flies mimic in dress various industrious or nest-building Bees, at whose expense they live in the manner of the Cuckoo. I found on the banks of the Amazons many of these Cuckoo Bees and Flies, which all wore the livery of working Bees peculiar to the country.

The instances of this kind of analogy most familiar to European entomologists are those of the European species of *Trochilium* (a genus of Moths), which strangely mimic various

* 'Rhopalocera Africæ Australis,' p. 21. Cape Town.

Bees, Wasps, and other Hymenopterous and Dipterous insects. The parallelism between these several forms and their geographical relations have not yet, I believe, been investigated. The resemblances seem to be more closely specific in tropical countries than in Europe; and I think it likely that the counterfeits in high latitudes may not always be found in company with their models. It is possible the geographical relations between the species concerned may have been disturbed by the great climatal and geological changes which have occurred in this part of the world since the date when they first came into existence.

Not only, however, are *Heliconidæ* the objects selected for imitation; some of them are themselves the imitators; in other words, they counterfeit each other, and this to a considerable extent. Species belonging to distinct genera have been confounded, owing to their being almost identical in colours and markings; in fact, many of them can scarcely be distinguished except by their generic characters. It is a most strange circumstance connected with this family, that its two sections, or subfamilies, have been mingled together by all authors, owing to the very close resemblance of many of their species. Analogies between the two subfamilies have been mistaken for affinities. It is sometimes difficult to understand in these cases which is the imitator and which the imitated. We have, however, generally a sure test in the one set exhibiting a departure from the normal style of colouring of their congeners, whilst the other are conformable to their generic types. The species of *Napeogenes* are, by this criterion, evidently all imitators of *Ithomiæ*; they are also rare insects, like the *Leptalides*. The mimetic species of *Heliconius* must be, for the same reason, imitators.

These imitative resemblances, of which hundreds of instances could be cited, are full of interest, and fill us with the greater astonishment the closer we investigate them; for some show a minute and palpably intentional likeness which is perfectly staggering. I have found that those features of the portrait are most attended to by nature which produce the most effective deception when the insects are seen in nature. The faithfulness of the resemblance, in many cases, is not so striking when they are seen in the cabinet. Although I had daily practice in insect-collecting for many years, and was always on my guard, I was constantly being deceived by them when in the woods. It may be asked, why are mimetic analogies so numerous and amazingly exact in insects, whilst so rare and vague in the higher animals*? The only answer that I can suggest is, that insects have perhaps attained a higher degree of specialization, after their type, than most other classes: this seems to be shown by the perfection of their adaptive structures and instincts. Their being more numerous and striking in tropical than in temperate countries is perhaps attributable to the more active competitive life, and the more rapid succession of their generations, in hot than in cold countries.

It is not difficult to divine the meaning or final cause of these analogies. When we

* Two instances of mimicry in birds, quite as wonderful as those between *Leptalis* and *Ithomia*, have just been communicated to me by my old travelling companion, Mr. A. R. Wallace. He has observed two species of *Oriolidæ* (perverted from the normal facies of the family) attendant on two species of *Meliphagidæ*, and mimicking them in the most curiously minute way in colours and in general figure. The associated pairs inhabit separate islands, as follows:—
—I. Bourou, *Mimeta* (*Oriolidæ*) *Bouroënsis*, *Tropidorhynchus* (*Meliphagidæ*), n. sp.; I. Ceram, *Mimeta Forstini*, *Tropidorhynchus subcarinatus*.

see a species of Moth which frequents flowers in the daytime wearing the appearance of a Wasp, we feel compelled to infer that the imitation is intended to protect the otherwise defenceless insect by deceiving insectivorous animals, which persecute the Moth, but avoid the Wasp. May not the Heliconide dress serve the same purpose to the *Leptalis*? Is it not probable, seeing the excessive abundance of the one species and the fewness of individuals of the other, that the Heliconide is free from the persecution to which the *Leptalis* is subjected?

I think it clear that the mutual resemblance in this and other cases cannot be entirely due to similarity of habits or the coincident adaptation of the two analogues to similar physical conditions. This is a very abstruse part of our subject; for I think the facts of similar variation in two already nearly allied forms do sometimes show that they have been affected in a similar way by physical conditions. A great number of insects are modified in one direction by a seaside habitat. I found, also, the general colours of many widely different species affected in a uniform way in the interior of the South American continent. But this does not produce the specific imitation of one species by another; it only prepares the way for it.

It is perhaps true that the causes (to be discussed presently) which produce a close or mimetic analogy cannot operate on forms which have not already a general resemblance, owing to similarity of habits, external conditions, or accidental coincidence. Species or groups which have this kind of resemblance to each other have been called by Dr. Collingwood recurrent animal forms. The English Bee-Moths owe the narrow and pointed shapes of their wings, which already approximate them to Bees, to their blood-relationship to the Hawk-Moth family. Their Bee-like size, form, and flight doubtless arise from their Bee-like habits. A close specific analogy between any one of these and a Bee, such as exists between the insects discussed in this memoir, could scarcely be due to an accidental resemblance like that between the Hawk-Moth and a Bee, or to similarity of habits. It would mean an adaptation of the Moth with especial reference to the Bee.

I believe, therefore, that the specific mimetic analogies exhibited in connexion with the *Heliconidæ* are adaptations—phenomena of precisely the same nature as those in which insects and other beings are assimilated in superficial appearance to the vegetable or inorganic substance on which, or amongst which, they live. The likeness of a Beetle or a Lizard to the bark of the tree on which it crawls cannot be explained as an identical result produced by a common cause acting on the tree and the animal.

Some of the imitations by insects of inanimate and living objects are very singular, and may be mentioned in this place. Many caterpillars of Moths, but sometimes the cases only which are manufactured and inhabited by the caterpillars, have a most deceptive likeness to dry twigs and other objects. Moths themselves very frequently resemble the bark on which they are found, or have wings coloured and veined like the fallen leaves on which they lie motionless. The accidental general resemblance between the shape of Moths' wings and leaves here gives nature the ground-work for much mimetic analogy. It has been pointed out by Rössler* that the Buff-tip Moth, when at rest, is intended to represent a broken piece of lichen-covered branch,

* In an article on resemblances between insects and vegetable substances (Wiener Entomol. Monatschrift, 1861,

—the coloured tips of these wings, when they are closed, resembling a section of the wood. Other Moths are deceptively like the excrement of birds on leaves. I met with a species of Phytophagous Beetle (*Chlamys pilula*) on the Amazons, which was undistinguishable by the eye from the dung of Caterpillars on foliage. These two latter cases of imitation should be carefully considered by those who would be inclined to think that the object of mimetic analogies in nature was simply variety, beauty, or ornament: nevertheless these are certainly attendants on the phenomena; some South-American *Cassidæ* resemble glittering drops of dew on the tips of leaves, owing to their burnished pearly gold colour. Some species of Longicorn Coleoptera (*Onychocerus scorpio* and *concentricus*) have precisely the colour and sculpture of the bark of the particular species of tree on which each is found. It is remarkable that other species of the same small group of *Longicornes* (*Phacellocera Buquetii, Cyclopeplus Batesii*) counterfeit, not inanimate objects, like their near kindred just cited, but other insects, in the same way as the *Leptalides* do the *Heliconidæ*.

Amongst the living objects mimicked by insects are the predacious species from which it is the interest of the mimickers to be concealed. Thus, the species of *Scaphura* (a genus of Crickets) in South America resemble in a wonderful manner different Sand Wasps of large size, which are constantly on the search for Crickets to provision their nests with. Another pretty Cricket, which I observed, was a good imitation of a Tiger Beetle *, and was always found on trees frequented by the Beetles (*Odontocheilæ*). There are endless instances of predacious insects being disguised by having similar shapes and colours to those of their prey; many Spiders are thus endowed: but some hunting Spiders mimic flower-buds, and station themselves motionless in the axils of leaves and other parts of plants to wait for their victims.

The most extraordinary instance of imitation I ever met with was that of a very large Caterpillar, which stretched itself from amidst the foliage of a tree which I was one day examining, and startled me by its resemblance to a small Snake. The first three segments behind the head were dilatable at the will of the insect, and had on each side a large black pupillated spot, which resembled the eye of the reptile: it was a poisonous or viperine species mimicked, and not an innocuous or colubrine Snake; this was proved by the imitation of keeled scales on the crown, which was produced by the recumbent feet, as the Caterpillar threw itself backwards. The Rev. Joseph Greene, to whom I gave a description, supposes the insect to have belonged to the family *Notodontidæ*, many of which have the habit of thus bending themselves. I carried off the Caterpillar, and alarmed every one in the village where I was then living, to whom I showed it. It unfortunately died before reaching the adult state.

p. 164). The author enumerates many very singular cases of mimicry; he also states his belief that the mimicry is intended to protect the insects from their enemies.

There is an interesting note, by the Rev. Joseph Greene, in the 'Zoologist,' 1856, p. 5073, on the autumn and winter Moths of England, whose colours are shown by the author to be adapted to the prevailing tints of nature in the season in which the species appear.

* A remarkable instance of deceptive analogy relating to a Cricket and a species of *Cicindela* is described by Westwood in Trans. Lin. Soc. vol. xviii. p. 419. In this memoir, Mr. Westwood has enumerated many curious cases of mimetic analogy.

I think it will be conceded that all these various kinds of imitative resemblances belong to the same class of phenomena, and are subject to the same explanation. The fact of one species mimicking an inanimate object, and another of an allied genus a living insect of another family, sufficiently proves this. I do not see how they differ from the adaptations of organs or instincts to the functions or objects they relate to. All are adaptations, either of the whole outward dress or of special parts, having in view the welfare of the creatures that possess them.

Every species in nature may be looked upon as maintaining its existence by virtue of some endowment enabling it to withstand the host of adverse circumstances by which it is surrounded. The means are of endless diversity. Some are provided with special organs of offence, others have passive means of holding their own in the battle of life. Great fecundity is generally of much avail, added to capabilities, active or passive, of wide dispersion; so that when the species is extirpated in one part of its area of distribution, the place is refilled by migration of individuals from another part. A great number have means of concealment from their enemies, of one sort or other. Many are enabled to escape extermination, or obtain subsistence, by disguises of various kinds: amongst these must be reckoned the adaptive resemblance of an otherwise defenceless species to one whose flourishing race shows that it enjoys peculiar advantages.

What advantages the *Heliconidæ* possess to make them so flourishing a group, and consequently the objects of so much mimetic resemblance, it is not easy to discover. There is nothing apparent in their structure or habits which could render them safe from persecution by the numerous insectivorous animals which are constantly on the watch in the same parts of the forest which they inhabit. It is probable they are unpalatable to insect enemies. Some of them (*Lycorea*, *Ituna*) have exsertible glands near the anus, which are protruded when the insects are roughly handled; it is well known that similar organs in other families (*Carabidæ*, *Staphylinidæ*) secrete fetid liquids or gases, and serve as a protection to the species. I have noticed also that recently killed specimens of Danaoid *Heliconidæ*, when set out to dry, were always less subject than other insects to be devoured by vermin. They have all a peculiar smell*. I never saw the flocks of slow-flying *Heliconidæ* in the woods persecuted by birds or Dragon-flies, to which they would have been easy prey; nor, when at rest on leaves, did they appear to be molested by Lizards or the predacious Flies of the family *Asilidæ*, which were very often seen pouncing on Butterflies of other families. If they owe their flourishing existence to this cause, it would be intelligible why the *Leptalidæ*, whose scanty number of individuals reveals a less protected condition, should be disguised in their dress, and thus share their immunity.

This explanation, however, would not apply to the imitation of Danaoid *Heliconidæ* by other species of the same subfamily. Moreover, there are several genera of other groups (*e. g.*, *Heliconius*, *Papilio*) which contain mimetic species side by side with species that are the objects of mimicry by members of other families, as will be seen by reference to the Table at p. 503. There is no reason to conclude that some of these possess the peculiar means of defence of the Danaoid *Heliconidæ*, whilst their near kindred are de-

* Mr. Wallace tells me the *Euplœæ* of the Eastern Archipelago have also this peculiar smell.

prived of them. It is not unreasonable to suppose that some species are taken by insectivorous animals, whilst others flying in company with them are avoided. I could not, from their excessive scarcity, ascertain on the spot that the *Leptalides* were thus picked out. I noticed, however, that other genera of their family (*Pieridæ*) were much persecuted. We have proof, in the case of Sand-Wasps, which provision their nests with insects, that a single species is very generally selected out of numbers, even of the same genus, existing in the same locality. I was quite convinced in the case of *Cerceris binodis* of South America, which destroys numbers of a *Megalostomis* (family *Clythridæ*), that the great rarity of the Beetle was owing to its serving as prey to the *Cerceris*. We cannot point out all the conditions of life of each species concerned in these mimetic analogies. All that we can say is, that some species show, by their great abundance in the adult state, that during this period, before they propagate their kind, they enjoy by some means immunity from effective persecution, and that it is therefore an advantage to others not so fortunate, and otherwise unprovided for, if they are so like as to be mistaken for them.

The process by which a mimetic analogy is brought about in nature is a problem which involves that of the origin of all species and all adaptations. What I have previously said regarding the variation of species, and the segregation of local races from variations, the change of species of *Heliconidæ* from one locality to another, and the probable vital necessity of their counterfeits which accompany them keeping to the exact imitation in each locality, has prepared the way to the explanation I have to give. In the cases of local variation of the *Heliconidæ*, there was nothing, as before remarked, very apparent in the conditions of the localities to show why one or more of the varieties should prevail in each over their kindred varieties. There was nothing to show plainly that any cause of the formation of local varieties existed, other than the direct action of physical conditions on the individuals, although this might be seen to be clearly incompetent to explain the occurrence of several varieties of the same species in one locality. We could only conclude, from the way in which the varieties occur in nature, as described in the case of *Mechanitis Polymnia*, that the local conditions favoured the increase of one or more varieties in a district at the expense of the others—the selected ones being different in different districts. What these conditions were, or have been, was not revealed by the facts. With the mimetic species *Leptalis Theonoë* the case is different. We see here a similar segregation of local forms to that of *Mechanitis Polymnia*; but we believe we know the conditions of life of the species, and find that they vary from one locality to another. The existence of the species, in each locality, is seen to depend on its form and colours, or *dress*, being assimilated to those of the *Ithomiæ* of the same district, which *Ithomiæ* are changed from place to place, such assimilation being apparently its only means of escaping extermination by insectivorous animals. Thus we have here the reason why local races are formed out of the natural variations of a species: the question then remains, how is this brought about?

The explanation of this seems to be quite clear on the theory of natural selection, as recently expounded by Mr. Darwin in the 'Origin of Species.' The local varieties or races cannot be supposed to have been formed by the direct action of physical conditions on the individuals, because, in limited districts where these conditions are the same, the most widely contrasted varieties are found existing together, and it is inexplicable how they could have produced the nice adaptations which these diverse varieties exhibit. All the varieties figured on Pl. LV. figs. 2, 7, 9, and on Pl. LVI. figs. 1, 2, 3, 4, 6, are found at St. Paulo, within a mile of each other, in the same humid forest. Neither can these adapted races, as before remarked, have originated in one generation by *sports* or a single act of variation in each case. It is clear, therefore, that some other active principle must be here at work to draw out, as it were, steadily in certain directions the suitable variations which arise, generation after generation, until forms have resulted which, like our races of *Leptalis Theonoë*, are considerably different from their parent as well as their sister forms. This principle can be no other than natural selection, the selecting agents being insectivorous animals, which gradually destroy those sports or varieties that are not sufficiently like *Ithomiæ* to deceive them. It would seem as though our *Leptalis* naturally produced simple varieties of a nature to resemble *Ithomiæ*; it is not always so, as is proved by many of them figured in the places above quoted. There is some general resemblance, it is true; and this is not purely accidental; for it is quite natural that the parent *Leptalis* should produce offspring varying in the direction of *Ithomiæ*, being itself similar to an *Ithomia*, and having inherited the property of varying in this manner through a long line of ancestors. We cannot ascertain, in this case, whether changed physical conditions have had any effect, quantitative or qualitative, on the variability of the species after migrating to a new district. At any rate, the existing varieties of our *Leptalis* show that the variations of *Leptalis* and *Ithomia* are not quite coincident, and that the agency of natural selection is required to bring the slowly forming race of one to resemble the other. I do not forget that at each step of selection the forms of *Leptalis* must have had sufficient resemblance to an *Ithomia* to lead to their preservation, or, at least, to prevent their complete extinction: as, however, the two analogues so much resemble each other at the commencement of the process, these steps would not be numerous. In many cases of mimetic resemblance, the mimicry is not so exact as in the *Leptalides*. This would show either that the imitator has only inherited its form from remote ancestors who were actively persecuted, the persecution having ceased during the career of its immediate ancestors; or it would show that the persecutor is not keen or rigid in its selection; a moderate degree of resemblance suffices to deceive it, and therefore the process halts at that point. I leave out of consideration all resemblances which can only be accidental, or which are resemblances of affinity.

If a mimetic species varies, some of its varieties must be more and some less faithful imitations of the object mimicked. According, therefore, to the closeness of its persecution by enemies, who seek the imitator, but avoid the imitated, will be its tendency to become an exact counterfeit,—the less perfect degrees of resemblance being, generation after generation, eliminated, and only the others left to propagate their kind. The actual state of *Leptalis Theonoë* is not the same in all of its three districts. A few varieties, or *sports*, are seen at Ega (65° W. long.) and St. Paulo (69° W. long.), namely, those figured Pl. LV. figs. 4, 5, 7, 8, and 9, which have an indeterminate resemblance. On the Cuparí

(55° W. long.) the resemblance is perfect (Pl. LV. fig. 1); and this is the only form of the *Leptalis* known in the locality. The varieties figured Pl. LVI. figs. 1, 2, 3, show different degrees of resemblance to *Ithomia Chrysodonia* (fig. 3 *a*); these, therefore, exhibit the selection in process. Thus, although we are unable to watch the process of formation of a new race as it occurs in time, we can see it, as it were, at one glance, by tracing the changes a species is simultaneously undergoing in different parts of the area of its distribution.

The fact of one of the forms of *Leptalis Theonoë*, namely *L. Lysinoë*, mimicking at Ega, not an *Ithomia*, but a flourishing species of another quite distinct family (*Stalachtis Duvalii*), shows that the object of the mimetic tendencies of the species is simply disguise, and that, the simple individual differences in that locality being originally in the direction, not of an *Ithomia*, but of another object equally well answering the purpose, selection operated in the direction of that other object. This point is well illustrated by the species of a small group of Longicorn Beetles already cited, some of which mimic a piece of bark, and others insects of another family—and by hunting Spiders, many of which wear the form of insects, and many that of inanimate objects amongst which they seek their prey.

When the persecution of a variable local form of our *Leptalis* is close or long continued, the indeterminate variations naturally become extinct; nothing then remains in that locality but the one exact counterfeit, whose exactness, it must be added, is henceforward kept up to the mark by the insect pairing necessarily with its exact counterpart, or breeding *in and in*. This is the condition of *Leptalis Theonoë* (Pl. LV. fig. 1) in its district; and it is the condition of all those numerous species of different orders which now appear fixed and distinct. When (as happens at St. Paulo, where a greater abundance of individuals and species, both of *Ithomia* and *Leptalis*, exists than in the locality of the last-named) many species have been in course of formation out of the varieties of one only, occasional intercrossing may have taken place; this would retard the process of segregation of the species, and, in fact, aid in producing the state of things (varieties and half-formed species) which I have already described as there existing.

In what way our *Leptalis* originally acquired the general form and colours of *Ithomiæ* I must leave undiscussed. We may conclude (if we are to reason at all from existing facts) that, as the antecedent forms of our races of *Leptalis* which are still undergoing change were themselves similar to *Ithomiæ*, the form has been inherited through a long line of ancestors, which have been more or less subjected to similar conditions. The instance of one of our forms leaving the *Ithomiæ* to mimic a species of another family may show us how a new line of mimetic analogy and gradual modification may have been originally opened.

Such, I conceive, is the only way in which the origin of mimetic species can be explained. I believe the case offers a most beautiful proof of the truth of the theory of natural selection. It also shows that a new adaptation, or the formation of a new species, is not effected by great and sudden change, but by numerous small steps of natural variation and selection. Some of the mutual resemblances of the *Heliconidæ* already mentioned seem not to be due to the adaptation of the one to the other, but rather, as they have a real affinity, the genera to which they belong being throughout very similar in colours and markings, and all equally flourishing, to the similar adaptation of all to the same local, probably inorganic, conditions. The selecting agent, which acts in each locality by destroying the variations unsuitable to the locality, would not in these cases be the same as in *Leptalis*; it may act, for anything we know, on the larvæ; in other respects, however, the same law of nature appears, namely, the selection of one or more distinct varieties by the elimination of intermediate gradations *. The conditions of life of these creatures are different in each locality where one or more separate local forms prevail, and those conditions are the selecting agents. With regard to the *Leptalides*, I believe we may be said to know these conditions. To exist at all in a given locality, our *Leptalis Theonoë* must wear a certain dress, and those of its varieties which do not come up to the mark are rigidly sacrificed. Our three sets of *Leptalides* may be compared to a variable flowering plant in the hands of a number of floriculturists, whose aims are different, each requiring a different colour of flower, and attaining his end by "roguing" or destroying all variations which depart from the standard.

It may be remarked that a mimetic species need not always be a rare one, although this is very generally the case; it may be highly prolific, or its persecution may be intermitted when the disguise is complete.

The operation of selecting agents, gradually and steadily bringing about the deceptive resemblance of a species to some other definite object, produces the impression of there being some innate principle in species which causes an advance of organization in a special direction. It seems as though the proper variation always arose in the species, and the mimicry were a predestined goal. This suggested the only other explanations that I have heard of, namely, that there may be an innate tendency in the organization to become modified in a given direction—or that the parent insect, being powerfully affected by the desire of concealment from the enemies of its race, may transmit peculiarities to its offspring that help it to become modified, and thus, in the course of many generations, the species becomes gradually assimilated to other forms or objects. On examination, however, these explanations are found to be untenable, and the appearances which suggest them illusory. Those who earnestly desire a rational explanation, must, I think, arrive at the conclusion that these apparently miraculous, but always

* Some of the close resemblances amongst the *Heliconidæ* themselves seem to be kept up by their varying in a precisely similar way. There is a very singular instance in three species of three different genera, *Melinæa*, *Mechanitis* (*Mothone*), and *Heliconius*, which are all, in East Peru, orange and black in colour, and in New Granada orange, black, and yellow. This seems to be a case of coincident, simple variation; for if three forms are quite alike in colours, it is conceivable that they may vary alike when placed under new conditions by migration. Our *Leptalides* have been shown not to vary precisely like their models; and therefore the case just quoted does not throw any difficulty in the way of the explanation I have given; but it is a very extraordinary one.

I have not thought it necessary to mention cases of close resemblances in insects which are only accidental, or which are explicable by the blood-relationship or affinity existing between the species which display them. Some orders of insects contain an almost infinite variety of forms, and it will not be wonderful, therefore, if species here and there be found to resemble each other, although inhabiting opposite parts of the earth, and belonging to widely different families. Such analogies are accidental, and can have nothing at all to do with the evidently intentional system of resemblances, carried on from place to place, which I have discussed. Some cosmopolitan families present very similar species in all parts of the earth; it can scarcely be necessary to say that close resemblances between New and Old World forms in these cases are resemblances of affinity, and not mimetic analogies.

beautiful and wonderful, mimetic resemblances, and therefore probably every other kind of adaptation in beings, are brought about by agencies similar to those we have here discussed.

[*Editor's Note:* Material has been omitted at this point.]

Part VII

MEASURING NATURAL SELECTION IN ACTION

Editor's Comments on Papers 21, 22, and 23

In both his books and letters Darwin stated his belief that natural selection operated too slowly in nature to be measured directly. However, the biologist E. B. Ford has reported that during the 1920s Leonard Darwin told him of a conversation that Leonard had had with his father, the great Charles Darwin:

> who expressed his belief that by choosing the right material it might be possible actually to detect evolutionary change taking place in natural conditions. For this purpose he said that long-continued observations and careful records would be needed, extending over a period which he estimated at perhaps fifty years. (Ford, 1964, pp. 298–299)

E. B. Ford went on to point out that

> As usual, Darwin was right, but on this occasion he was too pessimistic. The techniques of ecological genetics have made it possible not only to detect evolutionary changes in three or four generations of forms reproducing annually but to witness selection in operation and to observe and analyse its immediate effects. (Ford, 1964, p. 299)

Biologists began to employ statistics in an attempt to quantitatively measure selection taking place in natural conditions during the 1890s. W. F. R. Weldon (1860–1906), an English embryologist, became convinced that the operation of natural selection could be quantitatively measured using statistics as the result of reading Galton's *Natural Inheritance* (1889). He sought the help of the mathematician

Karl Pearson to develop statistical methods that could be used to ascertain whether natural selection was operating in a population. Weldon published his "Attempt to Measure the Death-rate Due to the Selective Destruction of *Carcinus moenas* with Respect to a Particular Dimension" in 1895 (Paper 21). The paper describes Weldon's measurement of the variation in frontal carapace breadth and in right dentary margin among 7000 immature and 1000 mature female shore crabs. His comparision of the measurements led him to conclude that natural selection was operating to produce a selective death rate—a death rate that increased as a function of the deviation from the mean with respect to frontal carapace breadth. In modern terminology, Weldon concluded that stabilizing, rather than directional, selection was operating.

Weldon carried out further researches on the selective death rate in shore crabs and presented his observations and conclusions in his 1898 "Presidential Address to the Zoological Section of the British Association" (Weldon, 1899). A comparison of the mean frontal breadth in terms of carapace length for male crabs collected in 1893, 1895, and 1898 led Weldon to conclude that the mean frontal breadth in this population of crabs was indeed changing. Weldon identified two environmental changes that were occurring in the habitats of the crabs—an increase in silt and in sewage. He conducted simulation experiments and reported that both an increase in china clay silt and an increase in putrid water produced a selective death rate that was higher among the male crabs having carapaces with larger frontal breadths. Weldon's attempts to quantitatively measure selection in nature and to simulate selection experimentally are important, not because he conclusively demonstrated the operation of a selective deathrate, but because he was the first to employ statistical techniques developed by Francis Galton to measure the effects of ongoing selection in nature. (See Cunningham, 1899, and Robson and Richards, 1936, for critical reviews of Weldon's research.)

Karl Pearson (1857-1936), the father of twentieth century statistics, developed statistical theory not for its own sake but for the study of problems in heredity and evolution. Weldon's 1895 research (Paper 21) served as the starting point for much of Pearson's research in developing and applying statistical techniques to more accurately measure the process of selection (Pearson, 1906, pp. 25-26). In his 1896 "Mathematical Contributions to the Theory of Evolution. III. Regression, Heredity and Panmixia" (excerpts presented as Paper 22), Pearson developed

> correlation, with special reference to problems in heredity, in which correlation and regression are defined in far greater gener-

> ality than previously and the theory of multivariate normal correlation is developed as a practical tool to a stage that left little to be added. (Eisenhart, 1974, p. 468)

In this paper Pearson coined two metaphors to describe different dimensions of selection. First, he defined selection that brings about a change in the mean and standard deviation of the variation-growth curve from one generation to the next as "secular selection" (now called directional selection). Second, he defined selection that maintains the same mean and standard deviation of the variation-growth curve from one generation to the next as "periodic selection" (now called stabilizing selection).

Pearson's investigations of ongoing natural selection in human populations led him to conclude that differential fertility appears to be countering natural selection in the case of civilized man (Paper 22, footnote, p. 259; Pearson, 1896, 1897, 1899). He considered differential fertility among individuals who obtained mates to be distinct from both sexual and natural selection, and yet so important in the human species that he coined a third metaphor "reproductive selection" to describe selection producing differential fertility among those who obtain a mate (Paper 22, pp. 258-259). Pearson borrowed data on height for couples and their offspring from Galton and made a number of comparisons, especially between individual parents and their adult offspring. Pearson's analyses of the upper middle-class families led him to conclude that reproduction in males is a function of height but that "it is impossible to more than *hazard* suggestions as to reproductive selection in relation to mother's height" (Paper 22, p. 275).

Pearson's 1896 paper is important for three reasons: (1) his development of statistical methods to make decisions concerning the direction and intensity of selection; (2) his application of these statistics to selection with respect to height; and (3) his conclusion that it is necessary to measure selective fertility among those individuals who survive and mate as well as to measure selective mortality (natural selection) and selective mating (Pearson's sexual selection).

A severe snow, sleet, and rain storm in Providence, Rhode Island, on February 1, 1898, caused many English sparrows to be blown down, 136 of which were brought to the anatomical laboratory at Brown University. These 136 sparrows were studied by Hermon C. Bumpus (1862-1943), who published the results of his statistical analyses of the natural selection generated by the storm as "The Elimination of the Unfit as Illustrated by the Introduced Sparrow, *Passer domesticus*" (Paper 22). He quantitatively measured nine physical characteristics of the downed sparrows and compared the measurements of the 72 birds that survived with those of the 64 birds that died. The selective mortality data published by Bumpus have been reanalyzed using more modern statistical techniques by Harris (1911), Calhoun (1947), Grant, (1972), Johnston et al., (1972), O'Donald (1973) and Lowther (1977). O'Donald (1973, p. 403) concluded that "The selection imposed on the sparrows of Providence, Rhode Island by the storm of 1898 was more intense than any which has since been observed acting on particular quantitative characters." Still higher selection intensities—the highest ever reported for any vertebrate population—have been recently published by Boag and Grant (1981) who analyzed the selective effect of the 1977 drought on the survival of Darwin's medium ground finch (*Geospiza fortis*) on the Galapagos island of Daphne Major.

REFERENCES

Boag, P. T., and P. R. Grant, 1981, Intense Natural Selection in a Population of Darwin's Finches (Geospizinae) in the Galapagos, *Science* **214:**82-85.

Calhoun, J. B., 1947, The Role of Temperature and Natural Selection in the Variations in Size of the English Sparrow in the United States, *Am. Nat.* **81:**203-228.

Cunningham, J. T., 1899, Professor Weldon's Evidence of the Operation of Natural Selection, *Nat. Sci.* **14:**38-45.

Eisenhart, C., 1974, Karl Pearson, in *Dictionary of Scientific Biography*, vol. 10, C. C. Gillespie, ed., Scribner's, New York, pp. 447-473.

Ford, E. B., 1964, *Ecological Genetics,* Methuen, London, 335p.

Galton, F., 1889, *Natural Inheritance,* MacMillan, London, 259p.

Grant, P. R., 1972, Centripetal Selection and the House Sparrow, *Syst. Zool.* **21:**23-30.

Harris, J. A., 1911, A Neglected Paper on Natural Selection in the English Sparrow, *Am. Nat.* **45:**314-318.

Johnston, R. F., D. M. Niles, and S. A. Rohwer, 1972, Hermon Bumpus and Natural Selection in the House Sparrow *Passer Domesticus, Evolution* **26:**20-31.

Lowther, P. E., 1977, Selection Intensity in North American House Sparrows (*Passer Domesticus*), Evolution **31:**649-656.

O'Donald, P., 1973, A Further Analysis of Bumpus' Data: The Intensity of Natural Selection, *Evolution* **27:**398-404.

Pearson, K., 1896, Contributions to the Mathematical Theory of Evolution. Note on Reproductive Selection, *Roy. Soc. London Proc.* **59:**301-305.

Pearson, K., 1897, *The Chances of Death and Other Studies in Evolution,* 2 vols., E. Arnold, London, 388p., 460p.

Pearson, K., 1899, Mathematical Contributions to the Theory of Evolution.—VI. Genetic (Reproductive) Selection: Inheritance of Fertility in Man and of Fecundity in Thoroughbred Racehorses, *Roy. Soc. London Philos. Trans.* **192:**257-330.

Pearson, K., 1906, Walter Frank Raphael Weldon. 1860-1906, *Biometrika* **5:**1-52.

Robson, G. C., and O. W. Richards, 1936, *The Variation of Animals in Nature,* Longmans, Green, London, 425p.

Weldon, W. F. R., 1899, Presidential Address to the Zoological Section of the British Association, *Br. Assoc. Adv. Sci. Rep. Bristol Meeting of 1898 Trans.,* pp. 887-902.

21

Reprinted from *Roy. Soc. (London) Proc.* **57**:360–379 (1895), by permission of The Royal Society, London

AN ATTEMPT TO MEASURE THE DEATH-RATE DUE TO THE SELECTIVE DESTRUCTION OF *CARCINUS MOENAS* WITH RESPECT TO A PARTICULAR DIMENSION

W. F. R. Weldon

February 28, 1895.

Sir JOHN EVANS, K.C.B., D.C.L., LL.D., Vice-President and Treasurer, in the Chair.

Professor Alexander Agassiz, who was elected a Foreign Member in 1891, was admitted into the Society.

A List of the Presents received was laid on the table, and thanks ordered for them.

This Meeting having been appointed by Council as a Meeting for Discussion, the following papers were taken as the subject of the discussion :—

I. Report of the Committee, consisting of Mr. Galton (Chairman), Mr. F. Darwin, Professor Macalister, Professor Meldola, Professor Poulton, and Professor Weldon, "for Conducting Statistical Inquiries into the Measurable Characteristics of Plants and Animals." Part I. "An Attempt to Measure the Death-rate due to the Selective Destruction of *Carcinus Mœnas* with respect to a Particular Dimension."—Drawn up for the Committee by Professor WELDON, F.R.S. Received November 20, 1894.

Among the material available for the purposes of the Committee was a sample of *Carcinus mœnas*, from Plymouth Sound, including a fairly large number of young females. The distribution of abnormalities in certain dimensions had already been determined for adult females from the same locality ('Roy. Soc. Proc.,' vol. 54, pp. 318—329); and it seemed worth while to compare the frequency of abnormalities in young individuals at various stages of growth with the frequency of the same abnormalities in adult life, so as to determine whether any evidence of selective destruction during growth could be discovered or not.

About 7000 females, varying in length from 7·00 to 13·95 mm., were chosen (at random, except as regards their size), and two dimensions were measured in each. The results were then compared with those of the corresponding measurements, made upon a sample of 1000 adult females from the same locality, which are recorded in the paper just referred to.

The dimensions chosen were:—(1) the "*frontal breadth*"—the distance in a straight line between the tips of the extra-orbital teeth of the carapace (from the point A, fig. 1, to the corresponding point on the opposite side); and (2) the "*right dentary margin*," measured in a straight line from the apex of the first to that of the last lateral tooth (from A to B, fig. 1). The "length" of each crab was taken as the length of the carapace, from the tip of the middle inter-orbital tooth to the posterior margin (from C to D, fig. 1). This is, of course, not the total length of the body; but the curvature and flexibility of the abdomen render an exact determination of the real body length very difficult.

FIG. 1.

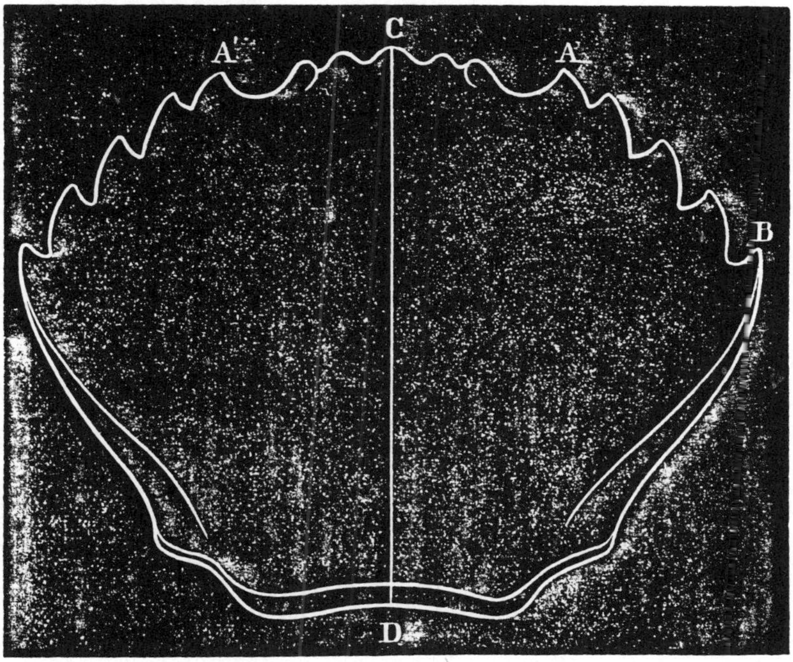

In order to compare the variability of a dimension in crabs whose carapace is only 7 mm. long with that of the corresponding dimension in adult crabs, whose carapace length is from 40—50 mm. or more, it is evidently necessary to adopt some method of picturing the crabs as of one standard size; and accordingly the measures obtained have always been expressed in terms of the carapace-length of the crab to which they belong, taken as 1000. The measurements were made by means of a screw, of 1 mm. pitch, carrying the object across the field of a microscope, and by means of graduations on the head of the screw the observations were recorded to the nearest hundredth of a millimetre. It is believed that the probable error of any observation is not much more than one hundredth of a millimetre. In order to minimise the effect of errors of observation, the results, after being expressed as fractions of the carapace-length, were sorted into groups, such that the measures in each group did not differ by more than 0·004 of the carapace-length, and all measures in the same group were treated as identical. The unit employed in tabulating the results was therefore 0·004 of the carapace-length; but in what follows the results are expressed, for the greater convenience of the reader, in thousandths of the carapace-length. It will be noticed that the principal effect of this alteration upon the results is to diminish their apparent regularity—an aberration of one unit of measurement appearing as four units in the tables below.

1. *Variation in Frontal Breadth.*

An initial difficulty in determining the error of distribution of frontal breadths about their mean in young crabs, arises from the great rapidity with which the mean itself changes during growth. The mean frontal breadth in the smallest specimens was found to be 853·14 thousandths of the carapace-length, while at maturity it is only 604·94 thousandths. The rate at which this change occurs can be gathered from the following table of the crabs measured, and the same result is graphically shown in fig. 2.

From this table it appears that the mean frontal breadth changes at such a rate that when the carapace-length has increased 0·2 mm., the frontal breadth has almost always diminished by less than four thousandths, that is to say, by less than one of the units of measurement here employed. For the purpose of the present investigation the mean was therefore considered stationary during every period of increase in size of not more than 0·2 mm., and the young crabs were accordingly sorted into groups, the individuals of each group differing by less than 0·2 mm. in respect of their carapace-length. The distribution of frontal breadths about the mean was then examined in each group separately.

As the difference in size between the largest and the smallest of the growing crabs was 7 mm., it follows that the material was divided into thirty-five groups. This subdivision of the material had great disadvantages, because, instead of a single group of over 7000 individuals, varying about the same mean, from which a fairly reliable indication of the law governing frequency of deviation might have been expected, the average number of individuals in any one of the available groups was only 200; and from so small a number of obser-

Table I.—Mean Frontal Breadth (F) expressed in thousandths of the Carapace-length, corresponding to various Carapace-lengths (C), together with the Number of Individuals on which each Determination is based.

C.	F.	Number.	C.	F.	Number.
7·1	853·14	159	10·7	798·01	225
7·3	852·43	186	10·9	794·96	162
7·5	850·89	172	11·1	792·14	222
7·7	844·27	142	11·3	789·26	218
7·9	844·22	132	11·5	789·26	230
8·1	837·13	224	11·7	786·07	211
8·3	835·41	219	11·9	784·53	225
8·5	830·08	214	12·1	782·42	224
8·7	826·80	207	12·3	780·92	226
8·9	823·75	214	12·5	778·39	219
9·1	821·26	191	12·7	772·76	183
9·3	818·33	205	12·9	771·62	233
9·5	815·89	214	13·1	770·36	131
9·7	811·60	195	13·3	769·86	162
9·9	809·95	226	13·5	767·70	158
10·1	809·27	245	13·7	762·51	201
10·3	803·21	253	13·9	763·47	211
10·5	800·53	232	(Adult)	(604·94)	(998)

[*Note.*—The carapace-length given in the table is the mean of all lengths included in each group. For example, the entry 7·1 includes all crabs measured in which the carapace-length was between 7·00 mm. and 7·19 mm., and so on.]

vations no satisfactory demonstration of the law of variation at any given moment of growth could be obtained. Nevertheless it was necessary, before proceeding further, to ascertain with some certainty what the law of variation through the whole series really was. The belief in which the work was undertaken was, that the law of variation would be found throughout to be that of the ordinary probability equation ; and this belief was tested in the following way:—In each of the thirty-five groups, the arithmetic mean of the frontal breadths, and the mean of all the deviations from it, were determined; and from the "mean error" found in this way the modulus of the probability function was calculated. Then, by calling the mean of each group zero, and expressing the deviations from the mean in terms of the modulus, a number of curves were obtained, in each of which the modulus was unity and the mean zero; a similar curve of adults was constructed, and the corresponding ordinates of all the thirty-six curves so obtained were added together. It is evident that, if the chance function really expresses the law of variation throughout the series, then the curve resulting from the treatment described will be a symmetrical probability curve of unit modulus. The actual

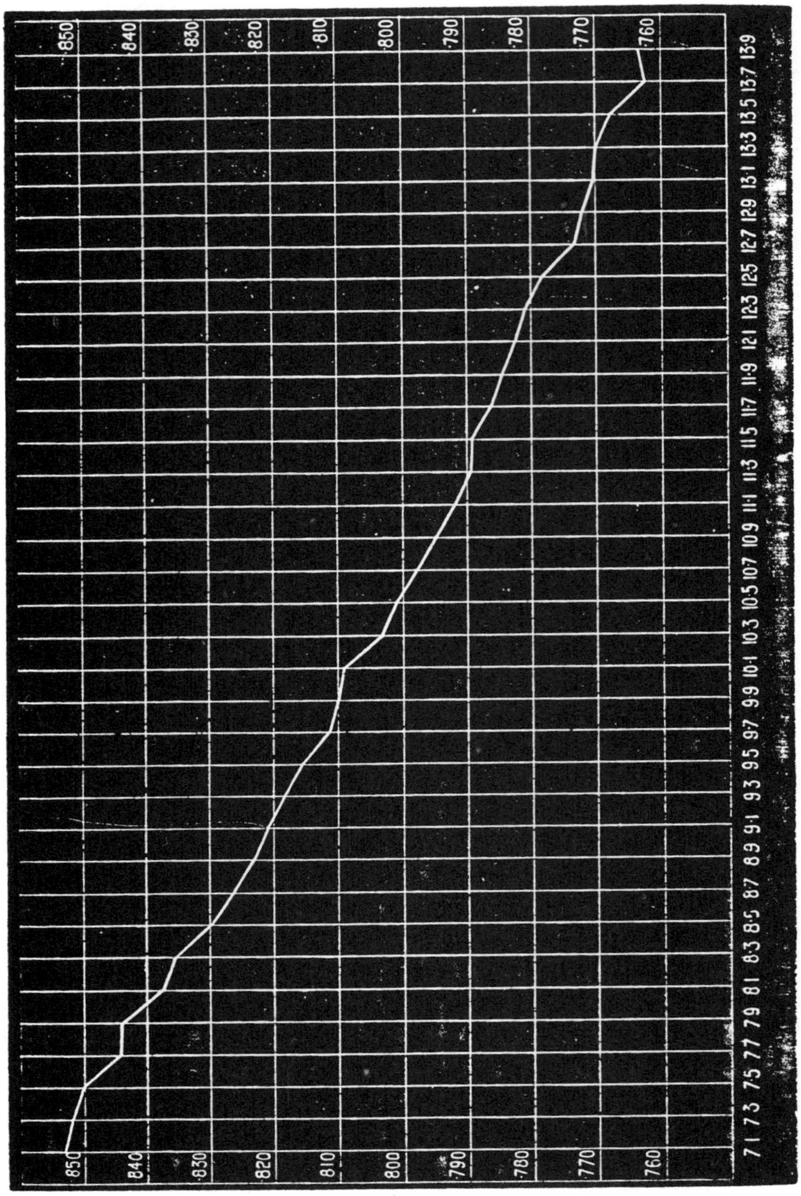

Fig. 2.—Diagram to illustrate the Change in the Mean Value of the Frontal Breadth with Growth in Carapace-length. Ordinates represent fractions of the Carapace-length. Abscissæ represent Carapace-length in millimetres.

curve obtained is plotted in Fig. 3, and the frequency of occurrence of every observed deviation is compared with that indicated by the tables of the probability integral in Table II. In spite of some discrepancies, the general agreement between the observed frequency of

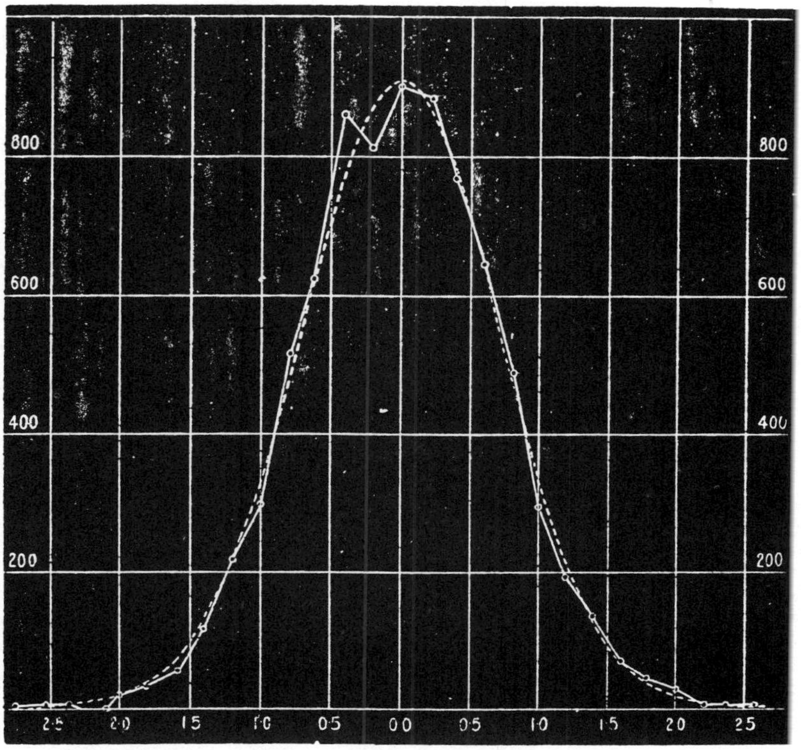

FIG. 3.—Distribution of Frontal Breadths in 8069 Female Crabs from Plymouth Sound, old and young. Deviations expressed in terms of the Modulus. The three cases of deviation greater than three times the Modulus are omitted.

deviations and that indicated by the probability integral is fairly close. The mean error of the observed curve is 0·5621, whereas it should be $1/\sqrt{\pi} = 0·5642$, the difference between the two figures being less than 0·5 per cent. The error of mean square is 0·7123, instead of 0·7071, a difference of less than 1 per cent. The sum of the squares of the positive deviations is 2115. The sum of the negative deviations is 1992. The total number of individuals of deviation more than 0·1 is 3593 on the positive, 3574 on the negative side, a difference of about one-half per cent.

On the whole it may be said that the result agrees with that given by the theory of probability as well as could be expected from the number of observations, and that the law of frequency of variation throughout the series may, as was hoped, be assumed to agree with the ordinary law of chance.

From the result so far obtained it followed that a determination of the quartile deviation, or any other of the constants of the pro-

Table II.—Frequency of all Observed Deviations from the Mean Frontal Breadth in 8069 Female Crabs, young and adult, from Plymouth. The Deviations expressed in terms of the Modulus.

Limits of deviations.	Mean deviation.	Observed frequency.	Theoretical frequency.
Over +3·29	+4·790	2	
From +3·10 to 3·29	+3·280	1	
" +2·90 " 3·09	—	0	
" +2·70 " 2·88	+2·880	1	12
" +2·50 " 2·69	+2·575	4	
" +2·30 " 2·49	+2·355	4	
" +2·10 " 2·29	+2·207	4	
" +1·90 " 2·09	+2·003	28	17
" +1·70 " 1·89	+1·788	46	36
" +1·50 " 1·69	+1·598	71	71
" +1·30 " 1·49	+1·395	137	129
" +1·10 " 1·29	+1·205	194	217
" +0·90 " 1·09	+1·006	295	336
" +0·70 " 0·89	+0·805	492	481
" +0·50 " 0·69	+0·589	649	635
" +0·30 " 0·49	+0·391	769	774
" +0·10 " 0·29	+0·191	896	872
" +0·09 " −0·09	−0·011	902	907
" −0·10 " 0·29	−0·213	814	872
" −0·30 " 0·49	−0·404	862	774
" −0·50 " 0·69	−0·611	626	635
" −0·70 " 0·89	−0·806	517	481
" −0·90 " 1·09	−1·000	299	336
" −1·10 " 1·29	−1·194	219	217
" −1·30 " 1·49	−1·403	118	129
" −1·50 " 1·69	−1·589	58	71
" −1·70 " 1·89	−1·808	33	36
" −1·90 " 2·09	−2·016	18	17
" −2·10 " 2·29	−2·210	1	
" −2·30 " 2·49	−2·368	4	
" −2·50 " 2·69	−2·520	3	12
" −2·70 " 2·89	−2·750	1	
—	—	—	
Over	−4·450	1	

bability equation, would be a trustworthy guide to the frequency of abnormalities at various periods of growth. But just as the individual groups were too small to allow of a determination of the law of abnormality in each, so they were too small to give trustworthy values of the quartile. The quartile deviation changes so slowly with growth, that it may without serious error be assumed to be constant during the period represented by 1 mm. of growth in carapace-length: that is, through the period covered by five of the groups into which the growing crabs were sorted. Therefore, after the quartile deviation had been determined in every group, the results were arranged in fives, and the mean of every consecutive five was taken as the quartile deviation through 1 mm. of growth. The results are shown in Table III.

Table III.—Quartile Deviation of Frontal Breadths (Q) for various Magnitudes of Carapace-length (C).

C.	Mean Q.
7·5	9·42
8·5	9·83*
9·5	9·51
10·5	9·58
11·5	10·25
12·5	10·79
13·5	10·09
(Adult)	(9·96)

The values here given are probably not very reliable, but they show that in the youngest individuals the quartile deviation is distinctly less than at maturity; that it increases with increase of size, until a time arrives when it is distinctly greater than in adult life; and that finally it diminishes again.

The initial features of this result,—the smallness of the quartile error at a young age, indicating relative infrequency of deviations, and the increase during growth, have been observed by Bowditch in the case of human stature. The result obtained by Dr. Bowditch and that here described are both simply confirmations of Darwin's statement, that many variations appear at a late period of development.

The initial increase in the quartile error may be attributed to the fact that average young produce upon the whole average adults, while animals which exhibit a deviation of known amount in the young state, exhibit on the whole a greater deviation with advancing age. If this view be the true one (and it is hoped that next year it may be possible to test it by observation of living crabs, which can be measured at various periods of growth), then, in a Plymouth crab, which is of unit deviation when its carapace is 7 mm. long, the most probable deviation when it has grown to be 12·5 mm. in length will be 10·79/9·42 = about 1·15 units. The probable error of this expectation is the expression of irregularities in the rate of growth, which cannot at present, for want of knowledge, be adequately discussed.

From the age represented by a length of 12·5 mm., the quartile

* Of the four very abnormal values shown in the table, three occurred in this group. They have been omitted in the determination of the quartile deviation, which would otherwise become 9·92.

error diminishes, and the parallel between the behaviour of the frontal breadth in Plymouth crabs and that seen by Bowditch in the stature of civilised human beings ceases to hold. The obvious suggestion by which to account for this seems to be that in the United States, where Bowditch made his observations, human beings are under conditions of such civilisation that there is considerable protection of the physically unfit; and that here, as in other civilised countries, any influences which might in a savage race produce selective destruction are reduced to a minimum, whereas in the case of the crabs such selective influences are active.

It is, of course, possible that the deviation of "abnormal" young may in each individual case first attain a maximum and then diminish with advancing age; if this is the case, we cannot know without experiment. In the absence of such experiment, the hypothesis may be provisionally adopted that the diminution in the frequency of individuals of given deviation is due to a selective destruction, and the consequences of this hypothesis will be examined.

Consider a population of crabs, measured at the time of their maximum variability, and suppose the distribution of deviations among the population to be accurately represented, for a particular organ, by a probability equation of modulus c_1. Now, let the number of individuals of deviation lying between $\pm a$ be represented by the area *abgd* (fig. 4); then, if $gd = 2a$ be small, compared with the observed range of variation, and $k_1 = \frac{\text{area } abgd}{2a}$, in other words, if k_1 be the height of the median ordinate BD of the generalised curve, then the whole number of individuals in the population will be $k_1 c_1 \sqrt{\pi}$.

Now, suppose any destruction, which acts unselectively with regard to the organ in question, to reduce the number of individuals whose deviation lies between $\pm a$, to *cdef*, and let the $\frac{\text{area } cdef}{2a} = k_2$, or the height of the median ordinate BD^1. Since this destruction is unselective, it will destroy an equal percentage of animals of every deviation, and will therefore not alter the modulus. The population will therefore be reduced to $k_2 c_1 \sqrt{\pi}$ in number. This unselective destruction cannot be directly measured.

The selective destruction is most simply conceived as follows:—

In fig. 4 let AD^1HE represent a curve of modulus c_1, and let $BD^1 = k_2$, so that the area of the whole curve $AD^1HE = k_2 c_1 \sqrt{\pi}$ represents the population left after unselective destruction has occurred. Then suppose the modulus to be reduced during growth to c_2, where c_2 is less than c_1, and let AD^1KE be a curve of modulus c_2. The minimum number of individuals which it is necessary to destroy, in order to affect this reduction in the modulus, is evidently represented by the

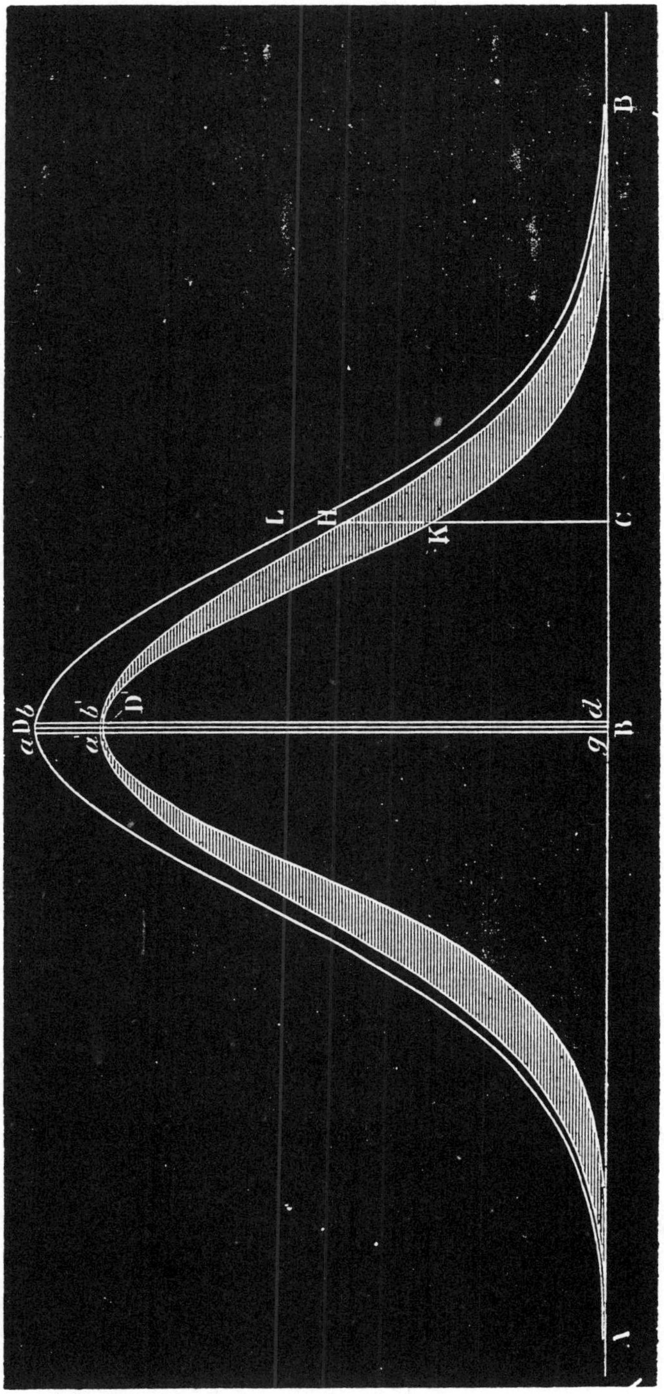

Fig. 4.

shaded area of the figure. The population after such destruction is $k_2c_2\sqrt{\pi}$ in number, and the shaded area represents a number of individuals equal to $k_2\sqrt{\pi}(c_1 - c_2)$, so that the ratio of animals selectively destroyed to animals which survive all unselective destruction is $\frac{c_1 - c_2}{c_1}$, a quantity which can be experimentally determined.

From the data given in Table III, this ratio, for frontal breadths of Plymouth crabs, becomes $\frac{10{\cdot}79 - 9{\cdot}96}{10{\cdot}79}$ or about 0·077, so that the hypothesis of selective destruction involves a death-rate of about 77 per thousand between the age corresponding to 12·5 mm. in carapace-length and maturity, as a consequence of deviation in frontal breadths, and in the group of structures, whatever these may be, which are directly correlated with it.

This total death-rate does not affect individuals of all deviations alike; an inspection of the figure will show that the death-rate is a function of the deviation, and that function is quite simply determined. Consider any ordinate Eg of the curve ABEC, and let its abscissa, DG, be of magnitude x; then the length of Eg is $k_2e^{-x^2/c_1^2}$ and the number of individuals of deviation between x and $x + dx$ is $k_2e^{-x^2/c^2}. dx$. In the same way, the height of the ordinate Fg is

$$k_2e^{-x^2/c_2^2},$$

and the number of individuals of abnormality within unit distance of x after selection is

$$k_2e^{-x^2/c_2^2} . dx.$$

The ratio between the number of animals of abnormality x which survive the unselective destruction and those which are selectively destroyed is therefore

$$\frac{\text{E}g - \text{EF}}{\text{E}g} = \frac{e^{-x^2/c_1^2} - e^{-x^2/c_2^2}}{e^{-x^2/c_1^2}} = 1 - e^{x^2(c_2^2 - c_1^2)/c_1^2c_2^2}$$

So that if g is the selective death-rate among animals of abnormality x, then that death-rate increases as x increases according to the equation

$$g = 1 - e^{-hx^2},$$

where h is the numerical value of $c_1^2 - c_2^2/c_1^2c_2^2$, c_1 and c_2 being the values of the modulus at the time of its maximum value and at maturity respectively.

For the frontal breadth of Plymouth crabs, the value of h is about 0·015; so that of the whole number of animals which attain the size 12·5 mm., having an abnormality x of their frontal breadth, the fraction destroyed as a consequence of this abnormality before reaching maturity is

$$1-e^{-0\cdot015x^2}.$$

It will, of course, be understood that little trust can be placed in the absolute numerical results which are here put forward; the point which seems worthy of confidence, and which if it be indeed a reality is of very great importance, is the form of the result. For by purely statistical methods, without making any assumption as to the functional importance of the frontal breadth, the time of life at which natural selection must be assumed to act, if it acts at all, has been determined, and the selective death-rate has been exhibited as a function of the abnormality, while a numerical estimate which is at least of the same order as the amount of the selective destruction has been obtained.

The method by which the result described has been arrived at is likely to be capable of application to a very considerable number of cases. Mathematically considered, the conditions which were at first assumed and then proved to obtain in the organ discussed are by no means general. It is necessary for the employment of this method that the variations should be distributed on each side of the mean with sensible symmetry, and that the position of minimum selective destruction should be sensibly coincident with the mean of the whole system. Such statistical information as is at present available leads to the belief that these conditions may be expected to hold for a large number of species, which are sensibly in equilibrium with their present surroundings, so that their mean character is sensibly the best, and the change of mean from generation to generation is at least very small. They cannot be expected to hold in cases of rapid change such as those induced artificially by selection under domestication, or naturally by rapid migration or other phenomena resulting in a rapid change of environment.

For the investigation of such rapid change, it would be necessary to treat the more general case, in which the chances of deviations of opposite sign are not sensibly symmetrical, and in which the mean is not necessarily the position of minimum destruction. The treatment of this case requires the help of a professional mathematician.

It will be well to mention here a curious indirect confirmation of the result just described, based on evidence derived from a quite different source.

An attempt has been made to show that physiological accidents of a kind leading to change in the length of a portion of the carapace affect a crab at a rate measured by the value of the quantity $1-e^{-hx^2}$. The symmetrical distribution of variations from the mean which has been shown, especially by Mr. Galton, to occur in dimensions of weight, length, muscular strength, and other characters of various organs in men, moths, sweet peas, and other things at various periods of life, made it seem probable that if selective destruction could be shown to occur in these cases, the expression for intensity of destruction, in terms of the deviation, would in all these cases be of the same form as that already arrived at. That is to say, the expression for the effect of physiological accidents of a number of different kinds, affecting a number of organisms in no way specially alike, is probably always of the same form. The question at once occurred, whether this expression might not be of general application, as a measure of the effect of physiological accidents upon the animal body.

The most convenient case in which to look for an answer to this question is the case of muscular tissue, in which the effect of accidents of stimulus can be readily measured. The recent paper of Cyboulski and Zanietowski (Pflüger's 'Archiv f. Physiologie,' Bd. 56, p. 45) gives an excellent series of data for determining the relation between energy of stimulus applied to a nerve, and effect upon the muscle, as measured by energy of external work performed in contraction. These observers give a large series of tables, in which the energy of stimulus, applied by discharging a condenser of known electrical capacity through a nerve, is given in one column, and in another is the work done by the muscle stimulated, measured by the height through which a known mass is lifted.

As is well known, the application of stimuli of less than a certain magnitude produces no muscular contraction; but if the maximum stimulus which can be applied without causing a contraction be reckoned as zero, the subsequent relation between stimulus and contraction does, in fact, agree very closely with that indicated by successive values of the quantity $1-e^{-hx^2}$.

In spite of the evident care and skill with which Cyboulski and Zanietowski have performed their experiments, their curves are slightly irregular. In order to minimise the effect of these slight irregularities, three of their results were treated in the following way:—In each system of observations the maximum subliminal stimulus was subtracted from the magnitude of the applied stimulus in each case; the three numbers representing the height of the muscle contraction for unit stimulus beyond this point in the three cases were added together; and so on throughout. The result is plotted in fig. 5, the height of the sum of three contractions being indicated by the ordinates of the points ⊙ ; the intensity of the corresponding stimulus, *minus* the subliminal stimulus, being measured along the abscissa.

The dotted curve is given by

$$g = 1-e^{-x^2/(8\cdot 15)_2},$$

8·15 being the "modulus" of the system of ordinates, determined from their moments about the axis of g. The coincidence between the two, rough as it is, is surely more than accidental!

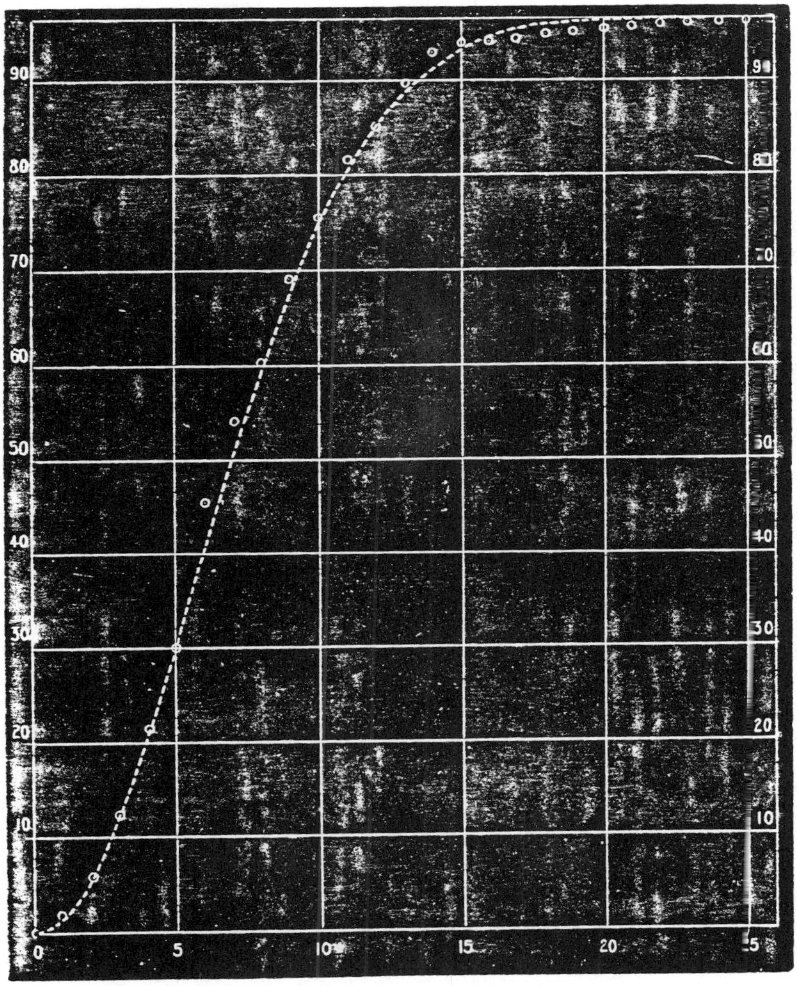

FIG. 5.—Sum of three Muscle-curves from Cyboulski and Zanietowski's paper. Each ordinate represents the sum of three muscle-contractions in millimetres: abscissæ represent stimulus applied to the nerve, expressed in ergs $\times 10^{-7}$, and reckoned from the close of the subliminal period.

The most interesting relation to be investigated in the light of this result would undoubtedly be the relation between sensation and stimulus in man; but existing data seem too imperfect to give any trustworthy result. It may be remarked that a few years ago measurements of the relation between muscle contraction and nerve stimulus, made with an imperfection comparable with that which characterises attempts to measure sensation, were held to obey a logarithmic law closely similar to the formula of Fechner and Weber.

II. *Variation in the Right Dentary Margin.*

The mean size of the right dentary margin was found to change, with increase of carapace-length, at such a rate as to render necessary the same subdivision of the material as that adopted in the case of frontal breadths. The change of mean will be gathered from Table IV, where it is seen that the change is slightly slower and less regular than in the frontal breadths, while its direction is reversed—the right dentary margin becoming larger, the frontal breadth smaller with increase of size.

Table IV.—Mean Length of Right Dentary Margin (D) expressed in thousandths of the Carapace-length (C) corresponding to various observed Carapace-lengths.

C.	D.	C.	D.
7·1	380·30	10·7	416·00
7·3	379·73	10·9	419·01
7·5	383·06	11·1	419·13
7·7	385·89	11·3	421·26
7·9	388·76	11·5	423·34
8·1	390·60	11·7	423·21
8·3	391·97	11·9	425·49
8·5	396·70	12·1	424·67
8·7	396·58	12·3	426·25
8·9	397·04	12·5	428·55
9·1	400·32	12·9	429·35
9·3	403·71	12·9	429·54
9·5	404·50	13·1	432·17
9·7	408·63	13·3	434·87
9·9	409·66	13·5	429·16
10·1	411·66	13·7	435·18
10·3	412·79	13·9	436·87
10·5	413·81	(Adult)	(495·14)

These observations were treated in the same way as those of frontal breadths; and the result of expressing the deviations from the mean in terms of the modulus of every group, and then summing deviations of corresponding magnitude, is shown in Table V, and graphically in fig. 6.

Table V.—Frequency of all observed Deviations from the Mean Length of the Right Dentary Margin in 8020 Female Crabs, Young and Adult, from Plymouth. Deviations in terms of the Modulus.

Limits of deviations.	Mean deviation.	Observed frequency.	Theoretical frequency.
	+5·530	1	
Over +3·20	+3·760	1	
	+3·340	1	
From +3·00 to 3·19	+3·120	1	2·76
„ +2·80 „ 2·99	+2·945	2	
„ +2·60 „ 2·79	+2·677	3	
„ +2·40 „ 2·59	+2·493	3	
„ +2·20 „ 2·39	+2·253	12	4·71
„ +2·00 „ 2·19	+2·050	9	11·29
„ +1·80 „ 1·99	+1·888	16	24·99
„ +1·60 „ 1·79	+1·705	45	51·14
„ +1·40 „ 1·59	+1·492	132	96·49
„ +1·20 „ 1·39	+1·276	155	168·30
„ +1·00 „ 1·19	+1·108	238	271·13
„ +0·80 „ 0·99	+0·897	375	403·41
„ +0·60 „ 0·79	+0·700	587	554·36
„ +0·40 „ 0·59	+0·493	735	703·61
„ +0·20 „ 0·39	+0·305	775	824·82
„ +0·00 „ 0·19	+0·112	939	893·04
„ −0·00 „ 0·19	−0·088	871	893·04
„ −0·20 „ 0·39	−0·270	833	824·82
„ −0·40 „ 0·59	−0·495	698	703·61
„ −0·60 „ 0·79	−0·739	553	554·36
„ −0·80 „ 0·99	−0·896	446	403·41
„ −1·00 „ 1·19	−1·093	240	271·13
„ −1·20 „ 1·39	−1·279	162	168·30
„ −1·40 „ 1·59	−1·479	89	96·49
„ −1·60 „ 1·79	−1·691	45	51·14
„ −1·80 „ 1·99	−1·932	18	24·99
„ −2·00 „ 2·19	−2·108	15	11·29
„ −2·20 „ 2·39	−2·295	6	4·71
„ −2·90 „ 2·59	−2·487	3	
„ −2·60 „ 2·79	−2·710	2	
„ −2·80 „ 2·99	−2·903	3	
	−3·330	2	2·76
	−4·180	1	
Over 3·00 ..	−4·830	1	
	−5·960	1	
	−7·030	1	

The symmetry of these results is fairly good, the number of positively abnormal individuals being 4030, the number of negatively abnormal 3990. The sum of the squares of the negative deviations is 2145·5; the sum of the squares of the positive deviations being 2099·9—a difference of about 2 per cent. This difference is greater than in the case of the frontal breadths; but a reference to the table will show that the removal of a single individual, namely, the indi-

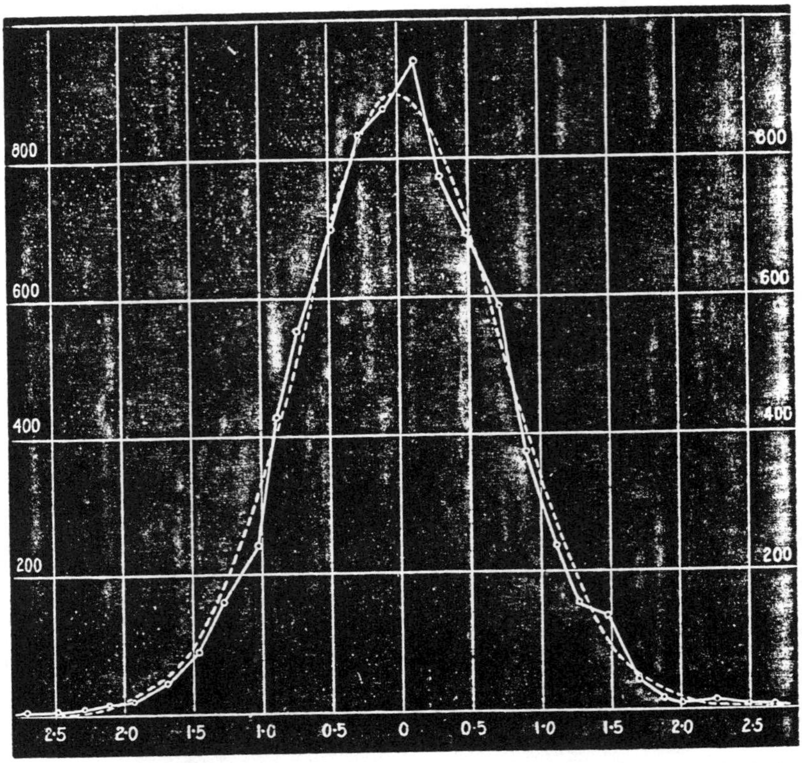

FIG. 6.—Deviations of 8020 measures of Right Dentary Margin in Female Crabs, old and young, from Plymouth Sound, expressed in terms of the Modulus. Eleven individuals of deviation greater than three times the Modulus are omitted.

vidual of deviation equal to —7 times the modulus, would make the sum of the positive and negative squares almost exactly equal.

The mean error of the whole system is 0·5688 instead of 0·5642, or nearly 1 per cent. too great. The error of mean squares is 0·7276 instead of 0·7077, or 2·8 per cent. too great.

From these values of the mean error and error of mean square, as well as from the presence of a deviation so great as seven times the modulus, it is evident that some cause has been at work, producing large abnormalities with a frequency greater than that indicated by the theory of chance. Reference to the table shows that deviations of more that 2·5 times the modulus do in fact occur twenty times, instead of five or six times, as they should do So that deviations of this magnitude occur about three and a half times too often.

The sporadic occurrence of considerable deviations, which do not obey the general law of frequency of variation, is a phenomenon which has been supposed by many naturalists to be of great importance in evolution, and the present case is therefore worthy of discussion.

The following suggestion is offered as an explanation of the large *negative* deviations. As shown in Fig. 1, there are normally five teeth in the dentary margin; but occasionally (in over 1 per cent. of individuals) only four teeth occur. The reduction in teeth may apparently be effected in various ways: sometimes it is impossible to say that one tooth rather than another is missing; and the case then resembles those cases of variation in the segmentation of a vertebral column, for example, recently discussed by Bateson ("Materials for the Study of Variation," *passim*, especially, however, p. 124). In other cases, the reduction appears to be effected by a process resembling the filling up of the interval between two teeth, so that the points of the teeth project only very slightly. A careful outline of a specimen exhibiting this condition is given in fig. 7. It is evident that

FIG. 7.

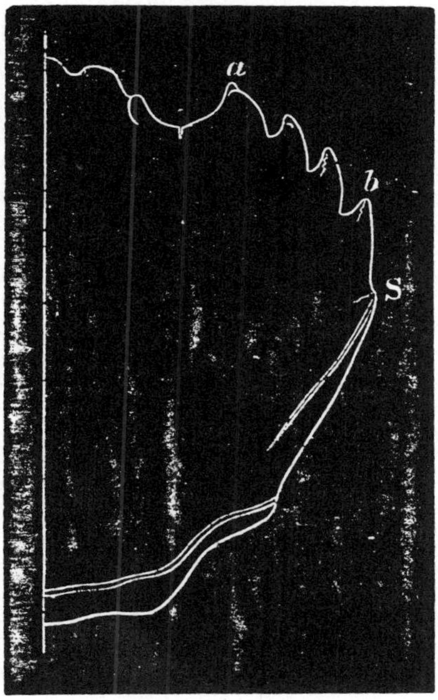

in this case the little tubercle S which indicates the position of the fifth tooth is the point from which the measurement should be taken; but if the obliteration of the fifth tooth had progressed but a little further, no indication of its presence would remain; and the dentary margin, measured from the tip of the first to the tip of the last visible tooth, would really have been measured from *a* to *b*. Such a measure would not be homologous with the rest, and ought not to be included in the scheme. But, since reduction in the number of teeth may take place in other ways, and since it is impossible in a given case to distinguish the manner in which it has occurred, the measurements were necessarily recorded.

Another frequent cause of disturbance is the breakage of the last tooth, followed by its regeneration. All cases of obviously recent injury were of course excluded; and for this reason the total number of individuals discussed is reduced from 8069 to 8020. But the selection of material was felt to be so dangerous a proceeding that all cases in which there was any doubt as to the occurrence of an injury were included. The wrongful inclusion of a dozen such cases would account for the excess of positive abnormalities: for it is evident that a breakage of the tip of the last tooth would increase the distance AB in fig. 1.

While, therefore, the observations admit of the interpretation that about once in 500 cases a "sport" of magnitude greater than that provided for by the theory of chance does regularly occur, the considerations which have been submitted make this interpretation at least doubtful.

The value of the probable error, as an indication of percentage abnormality, is diminished by the existence of the large deviations discussed; but the values obtained are of considerable interest: they are as follows:—

Table VI.—Mean Value of Quartile Deviation (Q) of Right Dentary Margin for various Lengths of Carapace (C).

C.	Mean Q.
7·5	8·44
8·5	8·08
9·5	9·36
10·5	8·23
11·5	8·16
12·5	8·05
13·5	8·68
(Adult)	(9·28)

It will be seen from this table that the error of distribution at the ages measured is always less than in adult life, except among crabs, whose carapace length is between 9 and 10 mm. Of the fourteen superfluous deviations of great magnitude, three occur in this group, and the result is a quite untrustworthy determination.

Evidently, therefore, in spite of the abnormally great frequency with which large deviations occur, the whole percentage of abnormalities, among crabs between 7 and 14 mm. in length, is less than it is in adult crabs; and there is a rough agreement between the result obtained from these measurements and that obtained by Bowditch from the measurements of human stature already referred to. So that among female crabs in Plymouth Sound, during the period of life to which these observations refer, there is no indication of any destructive agency which acts selectively upon the dentary margin. Whether such selective destruction occurs among males, or among females at a later period of life, is for the present an open question.

Variation in frontal breadth may, therefore, for the present be considered to be of more importance in the economy of female crabs than variation in the length of the dentary margin—a view which receives confirmation from the dimorphism already shown to exist ('Roy. Soc. Proc.,' vol. 54, p. 324) in the frontal breadth of crabs from Naples, while it is a striking justification of the accepted system of classification, in which the characters of the great groups into which the Brachyura are divided are almost entirely those associated with changes in this dimension.

In conclusion, an important feature of the method employed may be pointed out. The increase of death-rate, associated with a given abnormality of frontal breadth, has here been roughly determined; in the previous paper, already referred to, the effect of abnormality in this dimension upon several other organs of the body was determined; and by the method of that paper it would be possible to determine the effect of parental abnormality upon the offspring. These are all the data which are necessary, in order to determine the direction and rate of evolution; and they may be obtained without introducing any theory of the physiological function of the organs investigated. The advantage of eliminating from the problem of evolution ideas which must often, from the nature of the case, rest chiefly upon guess-work, need hardly be insisted upon.

22

Reprinted from pages 254-261, 269-278, 306-310, and 314-318 of *Roy. Soc. (London) Philos. Trans.* **187**(ser. A):253-318 (1896), by permission of The Royal Society, London

MATHEMATICAL CONTRIBUTIONS TO THE THEORY OF EVOLUTION.—III. REGRESSION, HEREDITY, AND PANMIXIA

Karl Pearson

[*Editor's Note:* The table of contents has been omitted.]

(1.) *Introductory.*

There are few branches of the Theory of Evolution which appear to the mathematical statistician so much in need of exact treatment as those of Regression, Heredity, and Panmixia. Round the notion of panmixia much obscurity has accumulated, owing to the want of precise definition and quantitative measurement. The problems of regression and heredity have been dealt with by Mr. FRANCIS GALTON in his epoch-making work on 'Natural Inheritance,' but, although he has shown exact methods of dealing, both experimentally and mathematically, with the problems of inheritance, it does not appear that mathematicians have hitherto developed his treatment, or that biologists and medical men have yet fully appreciated that he has really shown how many of the problems which perplex them may receive at any rate a partial answer. A considerable portion of the present memoir will be devoted to the expansion and fuller development of Mr. GALTON's ideas, particularly their application to the problem of *bi-parental inheritance*. At the same time I shall endeavour to point out how the results apply to some current biological and medical problems. In the first place, we must definitely free our minds, in the present state of our knowledge of the mechanism of inheritance and reproduction, of any hope of reaching a mathematical relation expressing the degree of correlation between individual parent and individual offspring.* The causes in any individual case of inheritance are far too complex to admit of exact treatment; and up to the present the classification of the circumstances under which greater or less degrees of correlation between special groups of parents and offspring may be expected has made but little progress. This is largely owing to a certain prevalence of almost metaphysical speculation as to the causes of heredity, which has usurped the place of that careful collection and elaborate experiment by which alone sufficient data might have been accumulated, with a view to ultimately narrowing and specialising the circumstances under which correlation was measured. We must proceed from inheritance in the mass to inheritance in narrower and narrwoer classes, rather than attempt to build up general rules on the observation of individual instances. Shortly, we must proceed by the method of statistics, rather than by the consideration of typical cases. It may seem discouraging to the medical practitioner, with the problem before him of inheritance in a particular family, to be told that nothing but averages, means, and probabilities with regard to large classes can as yet be scientifically dealt with; but the very nature of the distribution of variation, whether healthy or morbid, seems to indicate that we are dealing with that sphere of indefinitely numerous small causes, which in so many other instances has shown itself only amenable to the calculus of chance, and not to any analysis of the individual instance. On the other hand, the mathematical theory will be of assistance to the medical man by answering, *inter alia,* in its discussion of regression the problem as to the average effect upon the offspring of given degrees of morbid variation in the parents. It may enable the physician, in many cases, to state a belief based on a high degree of probability, if it offers no ground for dogma in individual cases.

One of the most noteworthy results of Mr. FRANCIS GALTON's researches is his discovery of the mode in which a population actually reproduces itself by regression and fraternal variation. It is with some expansion and fuller mathematical treatment of these ideas that this memoir commences.

* The physical and arithmetical statements of WEISMANN's "Theory of Germ Plasm" offer, so far as I have been able to interpret them, no sound basis for a quantitative theory of heredity in the mathematician's sense.

(2.) *Definitions.*

It is necessary to give definitions to several current biological conceptions, in order to introduce them into our mathematical analysis.

(*a.*) *Variation.*—If a curve be constructed, of which the ordinate y is such that $y\,\delta x$ measures the frequency with which an organ lying in size between x and $x + \delta x$, occurs in a considerable population (500 to 1000 or more), the constants which, for any particular organ for any particular animal determine the form of this curve, are termed the *constants of variation*, or more briefly, the variation of the given organ.

The assumption is made that the frequency is continuous, or that we really reach a curve. In the great majority of cases, where real statistical methods have been used, continuous curves (or, practically, polygons) have been found, and we shall assume this continuity to hold in all cases to which our formulæ are applied.

The size of the organ (x) which corresponds to the ordinate (y) through the centroid of the frequency curve, is termed the *mean;* the size of the organ, which corresponds to the ordinate bisecting the area of the frequency curve, is termed the *median;* the size of the organ corresponding to maximum frequency is termed the *mode.*

We assume, what may be considered as fairly established, that variation curves in zoometry, and more especially anthropometry, approximate closely to probability curves. When the variation curve has more than one mode, it may, as a rule, be resolved into simple probability curves, each with a single mode, and it may be even heterogeneous and require resolution, when only one mode is apparent.* These probability curves may be skew, and in this case the treatment of the problem of heredity involves a discussion of skew-correlation,† but in a very great range of cases the frequency is sufficiently closely given by the normal probability curve. Here the variation is defined by a single constant,‡ the standard deviation σ, and the equation to the curve is given by

$$y = \frac{N}{\sqrt{2\pi}\sigma} e^{-x^2/(2\sigma^2)},$$

and we shall confine our attention to such variation in the present memoir. The following assumption, therefore, lies at the basis of our present treatment of heredity. The variation of any organ in a sufficiently large population—which may be selected in any manner other than by this organ itself from a still larger population—is closely defined by a normal probability curve.

(*b.*) *Correlation.*—Two organs in the same individual, or in a connected pair of

* On resolution and skew variation, see 'Contributions to the Mathematical Theory of Evolution,' Memoirs I. and II., 'Phil. Trans.,' vols. 185 and 186.

† Dealt with in a memoir not yet published.

‡ Inheritance can be treated by single-constant variation in the case of most organs in human adults, but it could not be dealt with in like manner in the of case pedigree buttercups, see DE VRIES: 'Berichte der Deutschen Botanischen Gesellschaft,' 1894 and 1895.

individuals, are said to be correlated, when a series of the first organ of a definite size being selected, the mean of the sizes of the corresponding second organs is found to be a function of the size of the selected first organ. If the mean is independent of this size, the organs are said to be non-correlated. Correlation is defined mathematically by any constant, or series of constants, which determine the above function.

The word "organ" in the above definitions of variation and correlation must be understood to cover any measurable characteristic of an organism, and the word "size," its quantitative value.

(*c.*) *Natural Selection.*—This is of two kinds: *Secular Natural Selection* is measured by the changes due solely to mortality, in the mean and standard deviation of the variation-curve as we pass from one adult generation to the next. In statistical observations on man it is by no means easy—as we shall indicate later—to differentiate it from the effects of sexual selection, and of altered sanitary conditions.

Periodic Natural Selection may leave no trace of itself in the adult variation-curves of successive generations; it is measured by the changes due solely to mortality in the mean and standard deviation of the variation curves at successive stages of the *same* generation—*due allowance being made for the changes of the variation-constants due to growth.* In other words, if we watched a generation from birth to the adult stage, carefully preserving it from any form of selective mortality, such as arises from the struggle for existence, we should still find changes in the variation-constants due to the law of growth. If now the same generation be subjected to the struggle for existence, *i.e.*, placed in its natural surroundings, the variation-constants will differ from their values at the corresponding stages of the unselected growth. This difference is due to the selective mortality, *i.e.*, to natural selection. But this selective mortality may go on and still leave the variation-constants of the adult stage of each generation the same. In this case we speak of it as periodic natural selection. It repeats itself in each generation, but produces no secular change. It maintains an adult standard, but is not a factor of progressive evolution.

No estimate of periodic natural selection can be formed until the law of growth has been accurately ascertained by a series of observations on individuals. The influence of secular natural selection will be allowed for in our investigations by supposing the means and standard deviations of successive adult variation-curves to be not necessarily the same.*

(*d.*) *Sexual Selection.*—Sexual Selection† is of two kinds, due respectively to what

* Variation-curves for non-adult populations appear to be frequently skew. I propose in another paper to discuss the general law of selection on the basis of skew curves and with any arbitrary law of growth.

† I think DARWIN's view would be of the following kind. Let A be the most attractive female, *a* the most efficient male, Z the least attractive female and *z* the least efficient male. Then supposing only these four, *a* and *z* would both desire A with the result that (1) *a* would drive away *z* or (2) kill him. In the first case *z* would be free to mate with Z, but if he did so they would tend to produce a miserable

may be spoken of as individual and tribal taste. Tribal taste manifests itself in the preference of one sex as a whole for mating with members of the other sex having special characteristics, or to the rejection as mates by one sex of members of the other having special characteristics. The preference and rejection being in neither case absolute, but relative. This type of sexual selection, which may be spoken of as *preferential mating*, is measured by the differences in mean and standard deviation between the variation-curves for the whole adult population of one sex, and for the mated portion of it. For example, the mean height and mean variation in height of women generally are not identical, or are not necessarily identical with the mean height and mean variation in height of wives. Preferential mating may have reference to any organ or measurable characteristic of either sex.

Individual taste on the other hand does not denote the exclusion from mating of ·· section of the population of either sex. It is due to the preference of individuals with an organ or characteristic of given size for mates with the same or another organ or characteristic of a size, the average of which differs from the whole population average. This type of sexual selection which may be spoken of as *assortative mating* is measured mathematically by the coefficient of correlation between the two organs or characteristics in mated pairs.

It will be obvious that preferential mating and assortative mating are fundamental ideas to be quantitatively allowed for in any theory of heredity. Their action may often be in entirely opposite directions.*

(*e*.) *Reproductive Selection*.—One pair may produce more offspring than another, and in this manner give through heredity greater weight to their own characteristics. For example, the mean height of mothers is not identical, or is not necessarily identical with the mean height of wives, nor is the standard-deviation of fathers identical or necessarily identical with the standard-deviation of husbands. Further, the means and standard-deviations of mothers or fathers of sons may be different from those of mothers or fathers of daughters. The quantitative measure of reproductive selection is the correlation between the size of any selected organ in either male and female and their reproductivity, the reproductivity being measured by the number of their offspring in either sex or both sexes.

offspring fated to die out. Of course *a* might in certain cases after (1) mate with both A and Z. None of these possibilities corresponds exactly to what is described in this section as assortative mating, which in no way necessitates the exclusion from mating of *z*. *a* and *z* are not indeed competitors, but seeking different qualities in their mates. Thus, in man for example, the intellectual and non-intellectual might, and possibly do, sort themselves out in pairs, *i.e.*, there is a correlation between intellectual capacity of husband and wife.

* For example, preferential mating might lead in a highly social community to the rejection of consumptive mates, while assortative mating might, through localisation or community of habit, lead to considerable consumptive correlation. Thus sexual selection as a whole may influence in diverse ways the inheritance of the consumptive taint.

The importance of determining whether there is any correlation between reproductivity and a given organ of either parent appears to be great. For, if there be, it is not easy to understand how, even in the absence of both natural and sexual selection, a population can remain in a stable state. For example, suppose the mean father or the mean mother or both to be taller than the mean man or the mean woman or both, then this reproductive selection would appear to involve a gradual increase of height in the population in the same manner as selective breeding of animals by man might do. It is probable, therefore, that if reproductive selection be demonstrated by a finite value of the correlation constants, the instability of the population which results is partially or completely screened by natural selection.*

(*f.*) *Heredity.*—Given any organ in a parent and the same or any other organ in its offspring, the mathematical measure of heredity is the correlation of these organs for pairs of parent and offspring. If the organs be the same for parent and offspring, the heredity may be spoken of as *direct*, if they be different as *cross*. The word organ here must be taken to include any characteristic which can be quantitatively measured.

If the organs are not those of parent and offspring, but of any two individuals with a given degree of blood relationship, the correlation of the two organs will still be the proper measure of the strength of heredity for the given degree of relationship. *Cf.* § 6.

(*g.*) *Regression.*—Regression is a term which has been hitherto used to mark the amount of abnormality which falls on the average to the lot of offspring of parents of a given degree of abnormality. The mathematical measure of this special regression is the ratio of the mean deviation of offspring of selected parents from the mean of all offspring to the deviation of the selected parents from the mean of all parents. This may be further elucidated as follows:—Let parents, having an organ or characteristic of given deviation from the average or normal, be termed a "parentage," let the offspring of a parentage be termed a "fraternity." Then the *coefficient of regression* may be defined as the ratio of the mean deviation of the fraternity from the mean offspring to the deviation of the parentage from the mean parent. Both parentage and fraternity may be either male or female. It will be noted that we have so framed our definition of regression, that it marks the deviation of the fraternity from the filial and not the parental mean. We are thus able to allow for secular natural selection and reproductive selection. We shall see in the sequel that the coefficient of regression is a function of the variations in parents and offspring, and further of the correlations which define parental heredity and assortative mating. Further, as in heredity, the deviation or abnormality in parentage and fraternity may be measured with respect to the same or different organs; we have thus *direct* and *cross* regression.

From this special definition of regression in relation to parents and offspring, we

* I hope shortly to publish a paper on "Reproductive Selection in Man," and show how completely it appears to screen Natural Selection in the case of *civilised* man.

may pass to a general conception of regression. Let A and B be two correlated organs (variables or measurable characteristics) in the same or different individuals, and let the sub-group of organs B, corresponding to a sub-group of A with a definite value a, be extracted. Let the first of these sub-groups be termed an *array*, and the second a *type*. Then we define the coefficient of regression of the array on the type to be the ratio of the mean-deviation of the array from the mean B-organ to the deviation of the type a from the mean A-organ. The following are illustrations of types and arrays :—

Type.	Array.
Organ of given magnitude in—	Distribution of the correlated organs in—
Parent	Fraternity.
Offspring	Parentage.
Wife.	Male matage.
Husband	Female matage.
Given value of—	Distribution of correlated—
Height	Spans.
Cephalic index	Alveolar indices.
Barometric height	Heights at second station.
Local wages	Local pauper percentages.
Etc.	Etc.

It will be seen in the sequel that for the same pair of correlated organs or characteristics, the coefficient of regression is, if the law of frequency be the normal law, the same for the arrays corresponding to all types. But the coefficient is not the same when the type and array organs are interchanged, *e.g.*, the regression of husbands (male matage) on wives is not the same as the regression of wives (female matage) on husbands.

(*h.*) *Panmixia.*—Suppose that starting from a population of given mean and variation for any particular organ, secular natural selection of definite amount takes place for p generations and produces a population of another definite mean and variation for this same organ. Now suppose natural selection, whether periodic or secular, to be suspended for q generations, and sexual selection to be non-extant or negligible, then those members of the general population which were formerly weeded out, will now mix with all the other members of the population, and the results of interbreeding are spoken of as *panmixia.* The mathematical measure of the result on the given organ of panmixia acting for q generations is the change in mean and variation of the population with regard to that organ during these q generations. Should the mean and variation of the population tend with increase of q to approach the mean and variation of the population $p + q$ generations previously, panmixia may be said to reverse natural selection.

We have now defined the chief factors which will be dealt with in the present memoir, and shown how they are to be quantitatively measured. We shall now proceed to their mathematical analysis on the fundamental assumption that the variations with which we are about to deal obey the normal law of frequency.

(3.) *Correlation with special reference to the Problem of Heredity.*

(*a.*) *Historical.*—The fundamental theorems of correlation were for the first time and almost exhaustively discussed by BRAVAIS ('Analyse mathématique sur les probabilités des erreurs de situation d'un point.' Mémoires par divers Savans, T. IX., Paris, 1846, pp. 255-332) nearly half a century ago. He deals completely with the correlation of two and three variables. Forty years later Mr. J. D. HAMILTON DICKSON ('Proc. Roy. Soc.,' 1886, p. 63) dealt with a special problem proposed to him by Mr. GALTON, and reached on a somewhat narrow basis* some of BRAVAIS' results for correlation of two variables. Mr. GALTON at the same time introduced an improved notation which may be summed up in the 'GALTON function' or coefficient of correlation. This indeed appears in BRAVAIS' work, but a single symbol is not used for it. It will be found of great value in the present discussion. In 1892 Professor EDGEWORTH, also unconscious of BRAVAIS' memoir, dealt in a paper on 'Correlated Averages' with correlation for three variables ('Phil. Mag.' vol. 34, 1892, pp. 194-204.) He obtained results identical with BRAVAIS', although expressed in terms of 'GALTON'S functions.' He indicates also how the method may be extended to higher degrees of correlation. He starts by assuming a general form for the frequency of any complex of n organs each of given size. This form has been deduced on more or less legitimate assumptions by various writers. Several other authors, notably SCHOLS, DE FOREST and CZUBER, have dealt with the same topic, although little of first-class importance has been added to the researches of BRAVAIS. To Mr. GALTON alone is due the idea of applying these results—usually spoken of as "the laws of error in the position of a point in space"—to the problem of correlation in the theory of evolution.

The investigation of correlation which will now be given does not profess, except at certain stated points, to reach novel results. It endeavours, however, to reach the necessary fundamental formulæ with a clear statement of *what assumptions are really made*, and with special reference to what seems legitimate in the case of heredity.

* The coefficient of correlation was assumed to be the same for the arrays of all types, a result which really flows from the normal law of frequency.

[*Editor's Note:* Material has been omitted at this point.]

(*b.*) *Special case of Stature in Man.*—In order to get some idea of the nature of direct and cross inheritance, of assortative mating, &c., in man, I have, in conjunction with Professor W. F. R. WELDON, issued a circular and card appealing for help in collecting family measurements. We hope eventually to procure 1000–2000 families with data of height, span, and arm-length, but it may be many months, or even years, before sufficient material has been accumulated to allow of fairly definite statements being made. Meanwhile, Mr. GALTON, with his accustomed generosity, has placed at my disposal the family data on which his work on 'Natural Inheritance' was based. These data contain statistics with regard to one organ, *height*, for about 200 families. The number is not sufficiently great to make the probable error of quite small enough dimensions in several cases, and so allow of definite conclusions. The data do not offer, as those we are collecting, material for the treatment of cross as well as direct inheritance. Nevertheless, the drift of Mr. GALTON's statistics is in many cases obvious enough, and even in other cases, where the weight of the numerical results is not great, the conversion of our formulæ into numbers will still assist the reader to understand their significance, and serve to some extent for comparison when wider series of statistics are forthcoming.* Hence, in the numerical results of this paper, I wish more to draw attention to method than emphasise general laws. Mr. GALTON's families appear to have been drawn from the upper middle classes, and therefore any conclusions formed must not be hastily extended to the whole community.

* Only those who have attempted to get the measurements of, say 20 families, will appreciate the difficulty of the task of completing even 200 for one organ. Parents and children must be alive and fall within suitable limits of age; and what is more, their interest must be aroused.

The following tables give the chief results :—

TABLE I.—Variation.

Class.	Number.	Mean Height in inches.	Probable Error of M.H.	S.D. in inches.	Probable Error of S.D.
Males	683	69·215	·066	2·592	·047
Husbands	200	69·136	·126	2·628	·089
Sons	483	69·247	·081	2·617	·057
Fathers in general . . .	935	69·175	·055	2·501	·039
Fathers of sons	483	69·106	·071	2·325	·050
Fathers of daughters . .	452	69·248	·086	2·731	·061
Females	652	64·043	·061	2·325	·043
Wives	200	63·859	·110	2·303	·078
Daughters	452	64·118	·075	2·347	·053
Mothers in general . . .	935	64·099	·051	2·308	·036
Mothers of sons	483	64·054	·072	2·334	·051
Mothers of daughters . .	452	64·147	·072	2·274	·051

TABLE II.—Correlation.

Class.	Coefficient r.	Probable Error of r.
Husbands and wives	·0931	·0473
Fathers and sons	·3959	·0241
Fathers and daughters . . .	·3603	·0260
Mothers and sons	·3018	·0267
Mothers and daughters . . .	·2841	·0281

TABLE III.—Regression.

Class.	Coefficient of Regression.
Assortative Mating :—	
Husbands on wives	·1062
Wives on husbands	·0816
Inheritance :—	
Fathers on sons	·3517
Sons on fathers	·4456
Fathers on daughters	·4192
Daughters on fathers	·3096
Mothers on sons	·2692
Sons on mothers	·3384
Mothers on Daughters	·2753
Daughters on mothers	·2932

TABLE IV.—Variation in Selected Groups.

Class of Selected.	S.D. in inches.	S.D. in inches.	Unselected.
Matage :—			
Wives of selected husbands	2·293	2·303	All wives
Husbands of selected wives	2·617	2·628	All husbands
Parentage :—			
Fathers of selected sons	2·135	2·325	All fathers of sons
Fathers of selected daughters . . .	2·548	2·731	All fathers of daughters
Mothers of selected sons	2·225	2·334	All mothers of sons
Mothers of selected daughters . . .	2·180	2·274	All mothers of daughters
Fraternity :—			
Sons of selected fathers	2·403	2·617	All sons
Daughters of selected fathers . . .	2·189	2·347	All daughters
Sons of selected mothers	2·495	2·617	All sons
Daughters of selected mothers . . .	2·250	2·347	All daughters

TABLE V.—Sexual Ratio.

Class.	Ratio of Means.	Ratio of S. D.'s.	Ratio of V.'s.*
Husbands to wives	1·082	1·141	1·055
Males to females	1·081	1·115	1·032
Sons to daughters	1·080	1·115	1·032
Fathers to mothers . . , .	1·079	1·084	1·005

* V = the "coefficient of variation" or percentage of variation in organ = 100 S. D. ÷ (mean). See below.

N.B.—Mr. GALTON excluded from his calculations the larger families, but it seems to me that large families form an essential feature of the community. Two brothers are more likely to be two brothers of a large than of a small family, and, accordingly, large families ought to be given their proportionate weight. The whole problem, indeed, of reproductive selection turns upon the inclusion of large families.

Explanation of the Tables.—These tables were calculated in the following manner: Table I. A father or mother appears once for each child in this Table. The mean heights of each group were then calculated, as well as their standard deviations (S.D.) or deviations of mean square. The probable errors of the means and standard deviations were then found by means of the formulæ

$$p.e. \text{ of M.H.} = \cdot 674506 \times \text{S.D.}/\sqrt{n},$$
$$p.e. \text{ of S.D.} = \cdot 674506 \times \text{S.D.}/\sqrt{2n},$$

where n is the number of cases recorded in the second column of the Table.

To obtain Table II., tables of double entry* were formed for the class enumerated in the first column, *e.g.*, height of husband and height of wife as the variables x and y, and frequency of each pair of heights as z. From this table S (xy) was calculated by very laborious but straightforward arithmetic. This product moment was reduced to parallel axes (x', y') through the centroid of the system and r determined from the formula $r = \mathrm{S}(x'y')/n\sigma_1\sigma_2$ (see p. 265). The *p.e.* of r was then found from the formula on p. 266.

The coefficients of regression, in Table III., have the value $r\sigma_1/\sigma_2$, given on p. 267, where, if σ_1 be the standard deviation of A, and σ_2 of B, $r\sigma_1/\sigma_2$ is the regression of an A array on a B type, and $r\sigma_2/\sigma_1$, the regression of a B array on an A type.

In Table IV., the array is first stated and then the type; *e.g.*, in the first line the type is the husband of given height, the array the distribution of all wives of husbands of this height. The first S.D. is that of the array obtained from the formula S.D. $= \sigma_1\sqrt{1 - r^2}$, of p. 267, σ_1 being the second S.D. of Table IV., or the S.D. of the whole group from which the array has been extracted by selecting a particular value of the correlated group.

Table V. gives the ratio for corresponding groups of the two sexes of the constants given in Table I.

Now, a consideration of the probable errors recorded in Tables I and II. shows us that, in several cases, definite conclusions may be drawn, and in certain other cases very probable conclusions. In particular, the probable errors of the correlation coefficients of inheritance are sufficiently small to show that these coefficients give the chief features of heredity in the group and for the characteristic we are dealing with. We may note one or two special features.

(i.) *Natural Selection.*—We are dealing with two adult populations, and, therefore, should only expect to find traces of secular natural selection. The data, however, are not suited, either by their nature or number, to illustrate this point. There is a slight increase in height of sons over height of husbands, and a larger increase in height of daughters over height of mothers. Neither can be definitely asserted to be significant. Even if they were significant they might be accounted for by (a) shrinkage due to old age,† and (b) increased physical activity and exercise in the middle classes of the younger generation, especially daughters. If we turn from means to S.D.'s we see again an insignificant change in the range of variation of husbands and sons, the sons being slightly less variable than fathers. This result, were it necessary to account for it, would be more likely due to our having taken sons from a less general population than husbands—a point to be borne in mind

* It did not seem necessary to publish these tables, but the corresponding tables will be published when the fuller data for heredity in man, which I am at present collecting, are complete.

† In my own collection of data, several parents state that they are now shorter than they used to be. The shrinkage in the case of fathers of sons cannot be great in Mr. GALTON'S statistics, to judge by the means, unless we suppose a sensible regression in sons' stature.

when statistics of this kind are collected, and more than one son in a family is included. There is a more significant difference in the variation of wives and daughters. It is, however, in the opposite sense to what we may suppose would be produced by natural selection, or by the fact that we have drawn daughters from a less general population than wives. There is no definite evidence as to natural selection to be drawn from these results accordingly.

(ii.) *Sexual Selection.—(a.) Preferential Mating.*—We have no general populations to compare with those of husbands and wives. If we suppose the population stable, and treat sons and daughters as characteristic of the general unmarried population, husbands are not a significant selection from sons. Possibly the difference between the variation in daughters and wives might be accounted for by a distaste for very tall or very short wives in the middle classes. The difference is, however, not very significant, but it should be borne in mind in dealing with a larger range of statistics.

(*b.*) *Assortative Mating.*—Although the probable error (Table II.) is about half the coefficient of correlation, it is unlikely that the latter can be really zero, and although we must not lay very great stress on the actual value of *r*, still we are justified in considering that there is a definite amount of assortative mating with regard to height going on in the middle classes. It may be expressed by saying that wives 1″ taller than the mean will have on an average husbands ·11″ taller than the mean, and husbands 1″ taller than the mean, wives on an average ·08″ taller than the mean (Table III.). Table IV. shows us that the variation in matages would hardly be discoverable directly from our present range of statistics.*

(iii.) *Reproductive Selection.*—Although in the matter of means we cannot assert significance between the heights of males in general and fathers in particular, it is quite possible that such will reveal itself in more ample data. On the other hand, we see at once that fathers are definitely less variable than husbands, and fathers of sons remarkably less variable than fathers of daughters. Thus, while the height of a father is less closely related to his chances of having a daughter, any tendency to normality is of service in the chances of having a son. Reproductivity in males seems to be thus essentially correlated to height, and again, height to be potential in the question of male or female offspring.

An endeavour to directly calculate the correlation of reproductivity and height is

* Of course 200 couples give graphically nothing like a surface of correlation, nor can any section of it be taken as a fair normal curve. We assume *à priori* that 1000 couples would give a fair surface. This is practically what I have found for skull-measurements, 900 give an excellent curve, 50 a doubly, or even trebly, peaked polygon. None the less, sets of 50 skulls give means and S.D.'s in close accord. For example, in Professor FLINDERS PETRIE's newly discovered race, 50 male crania from T. and Q. graves give for cephalic index: Mean, 72·96, S.D., 2·82; while 53 male crania from General and B. graves give: Mean, 72·92, S.D., 2·95. The 103 crania together give: Mean, 72·938, S.D., 2·885, with a probable error of S.D. = ·29. The variation *curves* would not suggest any such close agreement at all. The constants, however, suffice to show the homogeneous character of the two sets of excavations.

frustrated by the obvious fact that size of adult families does not follow any approach to a normal distribution. Thus, I find in 205 adult families the following frequency:—

Title.	0	1	2	3	4	5	6	7	8	9	10	11	12	13	14	15
(1) Sons and daughters	..	32	22	23	31	23	19	18	18	8	5	5	..	..	..	1
(2) Sons only . .	25	43	46	42	30	10	5	3	..	..	1					
(3) Daughters only	25	56	44	34	21	13	5	3	3	1						

This Table shows the number of those families in which (1) the number of sons and daughters, (2) the number of sons only, (3) the number of daughters only correspond, with the title in the top line.

Now, although, as I propose to show later, the quantity, $r = \mathrm{S}\,(xy)/(n\sigma_1\sigma_2)$, is really a significant characteristic of correlation, just as σ_1 and σ_2 are significant for variation even in the case of skew frequency, still there is little to be gained by working it out in this particular case, where, the statistics being insufficient to accurately determine the skew law of frequency, we shall not be able to find what we want—the law of regression.*

But several points as to paternal reproductivity may be learnt from these families. In the first place, of the 25 families with no sons, the father in 5 cases only was below the mean, in 20 cases above the mean height. The mean height of fathers in general is 5′ 9″·17, but of sonless fathers is 5′ 11″·03. Of the 25 daughterless fathers, 14 are below and 11 above the mean height; the mean height of the daughterless father being 5′ 8″·71. Or, the same point may be emphasised in this way If short fathers be taken as those below 5′ 6″·5, and tall fathers as those above 5′ 11″·5, short fathers have 65 sons and tall fathers 67 sons. We should accordingly, with our proportion of sons and daughters, expect 61 daughters to short fathers and 63 to tall fathers, but we find short fathers with 73 and tall fathers with 81. This point in reproductive selection, that mediocre fathers have more tendency to sons and exceptional fathers to daughters, seems of considerable importance in relation to the prepotency of paternal inheritance. A similar point, but less emphatically significant, may be noted in the case of mothers. Mothers of daughters are less variable than mothers of sons. Without laying too great stress on statistics of so small a range and of one characteristic only, we may still suggest that it might be worth while to investigate whether the offspring of a mediocre parent and an abnormal parent do not tend to follow the sex of the mediocre parent.

* Much more complete statistics of size in families have recently been sent to me by Mr. F. HOWARD COLLINS. They give a remarkably smooth skew frequency distribution, thus demonstrating the need of the theory of skew correlation when we are dealing with reproductive selection. I propose to illustrate this in a memoir on skew correlation.

Finally, it is impossible to more than *hazard* suggestions as to reproductive selection in relation to mothers' height. It will be noticed that both mothers of sons and mothers of daughters are taller than wives, and, further, daughters, while taller than wives, are not so tall as mothers of daughters. Hence, while the difference in height of daughters and wives might be due to natural selection or improved physical training, it might also be accounted for by greater reproductivity as to daughters in tall women, *i.e.*, mothers of daughters taller than wives, and this tallness being transmitted in a lesser extent to daughters. This would be a case of secular change due to reproductive selection. The statistics are, however, too few to make the differences in the mean heights of wives, daughters and mothers, very definitely significant.

(iv.) *Inheritance.*—Mr. GALTON has concluded from his data that the coefficient of regression is ·3333 from father to son or from son to father, and by the assumption of the "midparent" has practically given the mother an equal prepotency with the father in heredity. The fuller theory developed in this paper does not seem in entire agreement with these conclusions. In the first place, the theory of uni-parental inheritance shows us that it is not the constancy of variation in two successive generations with which we have to deal, but the question whether sons have the same degree of variability as the "fathers of sons," and this must be definitely answered in the negative. Table II. shows us that there are undoubtedly significant differences in the coefficients of correlation, which may be summed up in the words *prepotency in heredity of the father.* It must be remembered that this is only for one characteristic, *height*, but in this characteristic both sons and daughters, on the average, take very considerably more after their father than after their mother. Turning to Table V., we see that the ratio of the mean heights of the two sexes, considered in three different classes, is practically the same, *i.e.*, 1·08, or 13 to 12, as Mr. GALTON has expressed it. Now, in Table III. we see that the coefficients of regression in paternal inheritance are not only sensibly greater than those of maternal inheritance, but, as these coefficients have to be multiplied by the *absolute* deviations of father or mother from their means to obtain the absolute deviations of offspring, and as these absolute deviations will be in the ratio of 13 to 12, there is a considerable further reduction to be made in comparing the strength of maternal with that of paternal heredity.

Thus it may be said that paternal heredity is to maternal heredity, in the case of sons, as ·4456 to $\cdot 3384 \times \frac{12}{13}$ or to ·3124, and in the case of daughters, $\cdot 3096 \times \frac{13}{12}$ or ·3354 to ·2932. Thus, while daughters inherit less from both their parents on the average than sons, both—and sons especially—take more after their father than their mother. The inferior inheritance of daughters may, to some extent, be counterbalanced by the law already noticed, that exceptional fathers have more often daughters than sons.

We may illustrate this by two examples—the regression of grandson on grandfather, and of great-grandson on great-grandfather when the inheritance is respectively through the male and female lines.

	Male line.	Female line.
Grandson on grandfather	·1986	·0385
Great-grandson on great-grandfather	·1048	·0307

In the first case, the strength of inheritance is more than double through the male; in the second, more than triple through the male what it is through the female line. Were this law of inheritance true, not only of stature, but of other physical, and especially of mental characteristics, some justification might be found for confining hereditary peerages initially given for merit to the male line. Meanwhile, it cannot be too strongly emphasized that the present results apply only to one organ, are based on comparatively few families drawn from a special class of the community, and thus stand in need of careful criticism in the light of ampler statistical material.

Another point already briefly referred to, which seems of significance, is the inequality of regression in the case of ascent and descent in the direct line. It may seem paradoxical to assert that sons are more like fathers than fathers are like sons, but the solution is bound up in the statement that fathers of sons are less variable than sons, or, in another form, that every son is not to the same degree a potential father. Similarly, the opposite paradox that fathers are, on the average, more like their daughters than daughters are like their fathers, finds its solution in the relatively great variability of fathers of daughters.

In Table IV. are tabulated alongside, in each case, the standard deviation for the corresponding general population, the standard deviations for inheritance from selected classes. Here again we see a general law for height, which deserves to be investigated for other organs, and for a variety of animals, namely, we breed "truer to the type," have less variability in offspring, if we breed from selected males rather than from selected females. We shall see later the effect of selecting both parents.

(*c.*) *On Further Relations between Correlation, Regression and Variability.* (i.) *The Coefficient of Variation V.*—In dealing with the comparative variation of men and women (or, indeed, very often of the two sexes of any animal), we have constantly to bear in mind that relative size influences not only the means but the deviations from the means. When dealing with absolute measurements, it is, of course, idle to compare the variation of the larger male organ directly with the variation of the smaller female organ. The same remark applies also to the comparison of large and small built races.

If the absolute measurements* have in the case of man to be on the average altered in the ratio of 13 to 12 to obtain those of the woman, if Mr. GALTON has gone so far as to replace any woman by an equivalent man on this basis, then, clearly, to compare

* The ratio 13 to 12 is not only true of stature, but approximately of several other organs, weight, brain-capacity, &c., &c.

deviations in man and woman, we must alter the deviations in the same ratio. Freeing ourselves from this particular ratio, we may take as a measure of variation the ratio of standard deviation to mean, or what is more convenient, this quantity multiplied by 100. We shall, accordingly, define V, the coefficient of variation, as the percentage variation in the mean, the standard deviation being treated as the total variation in the mean; since the p.e. = ·674,506 × S.D., V multiplied by ·674,506 may be called the "probable percentage variation." Of course, it does not follow because we have defined in this manner our "coefficient of variation," that this coefficient is really a significant quantity in the comparison of various races; it may be only a convenient mathematical expression, but I believe there is evidence to show that it is a more reliable test of "efficiency" in a race* than absolute variation. At present, however, we will merely adopt it as a convenient expression for a certain function, and proceed to examine its relation to correlation.

Let m_1, m_2 be the means of two correlated organs; σ_1, σ_2 their standard deviations; r their coefficient of correlation; V_1, V_2 their coefficients of variation; and R_1, R_2 the respective regressions for deviations d_2 and d_1 of the two organs.

Now

$$R_1 = r\frac{\sigma_1}{\sigma_2}d_2 = r\frac{V_1}{V_2} \times \frac{m_1 d_2}{m_2},$$

or

$$\frac{R_1}{m_1} = r\frac{V_1}{V_2} \times \frac{d_2}{m_2},$$

and similarly

$$\frac{R_2}{m_2} = r\frac{V_2}{V_1} \times \frac{d_1}{m_1}.$$

But we see that the amounts d_2/m_2 and d_1/m_1 are equally significant deviations in the case of the second and first organ, while the amounts R_1/m_1 and R_2/m_2 are equally significant regressions in the case of the first and second organ.†

It follows, therefore, that *the significances of the mutual regressions of the two organs are as the squares of their coefficients of variation.*

Hence inequality of coefficients of variation marks inequality of mutual regressions. Now coefficients of variation are rarely, if ever, equal for the same organ in corresponding classes of men and women. In dealing with male and female skull measurements for a great variety of races, this inequality is often very marked, and, therefore, differences of variation tell, especially in mutual regression in the case of sexual selection and inheritance from the opposite sex. They are sufficient, I think, to preclude Mr. GALTON's theory of the mid-parent from being considered as more than a first approximation.

* By "race efficiency," I would denote stability, combined with capacity to play a part in the history of civilization. I hope later to publish details of variation, especially in skull measurements of different races of man, the data of which I have been for some years reducing.

† For example, 1″ and $\frac{13}{13}$″ I term equally significant deviations or regressions in the stature of man and woman, and 1″ and $\frac{13}{12}$″ in the stature of woman and man.

[*Editor's Note:* Material has been omitted at this point.]

Thus we conclude that

$$\zeta = r \frac{\sigma_1}{\sigma_2} h_2 + r \frac{\sigma_2}{\sigma_3} h_3, \qquad \sigma'^2 = \sigma_1^2 (1 - r^2 - r^2),$$

the formulæ for biparental inheritance with equal parental correlation, and no assortative mating. The analysis for unequal parental correlation and assortative mating follows the same lines, is far more lengthy, but leads to the same result, *i.e.*, no gain by selection of the same amount, oft repeated.

(*d.*) *Secular Natural Selection and Steady Selection. Focus of Regression.*—We thus see that, on the theory with which we are concerned, a knowledge of the ancestry beyond the parents in no way alters our judgment as to the size of organ or degree of characteristic probable in the offspring, nor its variability.* *An exceptional father is as likely to have exceptional children if he comes of a mediocre stock as if he comes of an exceptional stock.* The value of ζ will be no greater nor the value of Σ_1 less if the parents have been selected for p generations than if they have been selected for one only. This result seems to me somewhat surprising, but I cannot see how it is to be escaped so long as we assume the normal distribution of frequency, which appears in so many cases to be a close approximation to fact. It is of course possible

* This seems specially noteworthy; it would seem natural to suppose that the offspring of a long selected stock would be less variable than those of one just started—that the offspring of a gradually created variety would be more stable than those so to speak of a sport. It appears not.

that in some manner repeated selection causes a progression of the "focus of regression," by which term I would understand the mean of the general population from which selection has originally taken place. I have been very careful so far not to hazard any statement with regard to this focus of regression. I have measured only the amount by which the offspring of exceptional parents diverge, not from the mean of the parental population but from the mean of the offspring population. In this manner our formulæ allowed for the play of secular natural selection. It is quite true that the word "regression" thus loses its accustomary meaning, which it can only bear if the population be stable and the means of two generations sensibly identical; this is the case for which, I think, the word was introduced by Mr. GALTON.* The sense given to it in the present paper is accordingly a technical one; as already defined, it is the ratio of the mean deviation of the offspring of a selected parentage to the deviation in the parent which characterises the selection, the deviations in offspring and parents being respectively measured from the means of the corresponding general populations. Now here, at the very outset of our consideration of panmixia arises a very real difficulty, which is vital for the whole theory of evolution by natural selection. According to Mr. GALTON the population being stable, or no secular natural selection or reproductive selection taking place, there is a regression of the offspring of selected parents towards the mean of a certain general population, and the "grandchildren" also regress to the same mean. We shall see then that unless correlation is perfect $\left(r\,\dfrac{\sigma_2}{\sigma_1} = 1\right)$ no amount of continued selection would suffice to prevent a race from regressing to an original general population when that selection was suspended. Panmixia in the sense of its most ardent supporters would be demonstrated. But the difficulty is not the establishment of panmixia, but as to what is to be considered the "original general population." On the theory of evolution by natural selection that general population has itself been produced by a series of selections, and selections probably affecting its mean as well as its standard-deviation, hence how is it possible to pick out any particular stage of general population as the "focus of regression," and assert that regression of the offspring of parents now selected takes place towards that stage of evolution? Where is the focus of regression to be placed for the profile angle of man? About 80° to 90° or nearer the 40° to 70° of the anthropoid apes? The further back the better for those who believe that suspension and reversal of natural selection are identical, but no manipulating whatever of the human mortality tables would allow for a "focus of regression" very considerably below that of the current general population. Hence it would seem essential that successive selections must connote some progression of the focus of regression. This progression may be continuous with continuous natural selection, or it may take place by starts and leaps, as

* We can at once restore the true notion of regression, as Mr. GALTON points out to me, by measuring each organ or characteristic in terms of its own standard-deviation. It will then be a coefficient of correlation and a proper fraction.

indicated in Mr. GALTON's idea of organic stability. In either case panmixia would only carry back the mean to the current focus of regression, and so be a very minute reversal of natural selection.

What our theory really shows is a regression of the offspring of selected parents towards the mean of general offspring. This latter mean, supposing no secular natural selection, can, it seems to me, only be determined by experiment. It can hardly agree with the general parental mean, if the parents themselves are the product of natural selection. On the other hand, the statistics actually obtained for stable, or sensibly stable, populations seem to mark a focus of regression close to the mean of the current population, and, therefore, a progression of the focus due to past selection.* Meanwhile, till experiment has settled how continuous selection affects the focus of regression, we may see whither extreme hypotheses lead us. Such are:

(1.) The focus of regression remains stable during selection.

(2.) The focus of regression is the mean of the population from which parents have been selected.

(*c.*) *Focus of Regression Stable during Selection.*

(i.) *Steady Selection cannot be Secular or Produce Truer Breeding.*—We have seen that on this hypothesis ancestry, as distinct from immediate parentage, is indifferent. Thus, in the case of parthenogenetic reproduction, or of sexual reproduction with one parent selected, we have seen that one selection leads to the distribution (Cor. 1, p 301):

$$\zeta = r_{01} \frac{\sigma_0}{\sigma_1} h_1, \quad \Sigma^2 = \sigma_0^2 (1 - r_{01}^2) + r_{01}^2 \frac{\sigma_0^2}{\sigma_1^2} s_1^2,$$

and if out of this we again select a parentage, defined by h_1 and s_1, we shall obtain the same distribution of offspring, and this however often the process be repeated. We must increase the divergence (h_1) of the selected from the general population or its concentration ($1/s_1$) or both, if we require any progressive effect from continual selection. The same remarks apply to bi-parental selection (m_1, m_2, s_2, s_3, ρ_{12}). Persistent selection only suffices to keep the mean and variation at a definite distance from those of the general population. Or, on the hypothesis of a stationary focus of regression, we conclude that *steady selection, however long it persists, can only be periodic and not secular.*

This point seems of such importance that it may be best to illustrate it by an example drawn from our Table I. and Table III. The mean height of fathers being about 69″·2, the regression of the average sons of fathers of 6′ in height is about 1″·25, or the average height of sons of a 6′ fatherhood = 70″·45. Hence, if we select fathers forming a normal distribution of any standard round 6′, we shall have a normal distribution of sons round 70″·45. If we select a second parentage from

* The determination of the focus of regression for some organ in selected domestic ducks for several generations and comparison with the means for wild and general domestic ducks would seem a possibility.

those taller sons averaging 6′, their sons will still only average 70″·45, and, however long we persist in this process of selection, we shall produce no secular change; the population will remain after the pth selection just where it was both as to mean and variation after the first. The only way to produce a secular change is to continually increase the standard of the selected (or to alter the focus of regression). No steady selection would appear to produce "truer breeding."

(ii.) *Panmixia and Uni-parental Regression.*—Continual selection of the same magnitude for p generations, merely giving us the same mean and variation, we may now ask what would be the effect of suspending natural selection for q generations.

Take first the case of parthenogenetic reproduction, or that of uni-parental regression. The first parentage after suspension of natural selection will have $m_2\, r_3\, \sigma_1/\sigma_2$ for its mean, and $\sqrt{\{\sigma_1^2 (1 - r_3^2) + s_2^2\, r_3^2\, \sigma_1^2/\sigma_2^2\}}$ for its standard-deviation. Successive parentages can be found by substituting these values successively in themselves for the quantities m_2 and s_2. We find at once that after q generations of suspended selection the mean of the population will differ from the focus of regression by

$$m_2 (r_3\, \sigma_1/\sigma_2)^q,$$

and the standard-deviation will be given by

$$\Sigma_q^2 = \sigma_1^2 (1 - r_3^2) \frac{1 - (r_3\, \sigma_1/\sigma_2)^{2q-2}}{1 - (r_3\, \sigma_1/\sigma_2)^2} + (r_3\, \sigma_1/\sigma_2)^{2q-2}\, s_1^2.$$

Now if the population simply repeat itself without any natural selection (if there be no reproductive selection at work) $\sigma_1 = \sigma_2$, and in most cases I have come across $r_3\, \sigma_1/\sigma_2$ is a fraction. Hence, as q is indefinitely increased $m_2 (r_3\, \sigma_1/\sigma_2)^q$ becomes indefinitely small, and $\Sigma_q^2 = \sigma_1^2$, or $= \sigma_1^2 \dfrac{1 - r_3^2}{1 - (r_3\, \sigma_1/\sigma_2)^2}$, if σ_1 be not equal to σ_2.

We see, therefore, that both as to mean and variation the population with suspended natural selection tends to rapidly regress to the general population from which it was selected. This is still true if there has been a continuous secular, as distinguished from a periodic natural selection, for we have only to suppose m_2 and s_2 to be the final result of such selection. If then the focus of regression does not progress with continuous selection, all that has been asserted as to the effect of suspended natural selection holds, at least so far as concerns a return to the condition of things which prevailed when the focus of regression was the mean of the general population. But unfortunately the advocates of panmixia want more than this, namely, either an indefinite regression of the focus of regression itself, or to place it, if steady, at an indefinitely distant point. The first result would be perfectly parallel with our second hypothesis—a progression of the focus of regression,—but would demand rather a reversal than a suspension of natural selection. The second result seems quite inconsistent with any statistics of successive generations yet taken; it demands a mortality due to natural selection, which its propounders have hardly appreciated.

[*Editor's Note:* Material has been omitted at this point.]

(f.) *Progression of the Focus of Regression with Natural Selection.*

(i.) *General Remarks on Regression and Fixedness of Character.*—Our first hypothesis certainly favours the general views of those who support the doctrine of panmixia, although to be quite consistent they must :

(i.) Place the focus of regression back at the zero size of an organ or the zero degree of intensity of a characteristic.

(ii.) Assume much nearer approximation to unity in their coefficients of regression than any measurement as yet suggests, or

(iii.) Demand a far higher mortality of *periodic* natural selection than has anywhere as yet been demonstrated.

Professor WEISMANN has no difficulty, apparently, about (i.): "As soon as natural selection ceases to operate upon any character, structural or functional, it begins to *disappear.*" ("Essays on Heredity," 1889, p. 90.) He talks of functionless organs losing in size with the suspension of natural selection "until the last remnant finally *disappears*" (*ibid.*, p. 292), while "the disposition of the tail to become rudimentary, in cats and dogs, may be explained in the simplest way, by the process which I have formerly called panmixia," *i.e.*, suspension of natural selection (*ibid.*, p. 430). This explanation "in the simplest way" fails entirely to say whether (ii.) or (iii.) is to be accepted after assuming the truth of (i.). What is quite clear is that in the only case where either the coefficients of regression or the mortality can at present be even approximately stated neither (ii.) nor (iii.) hold. Fox-terriers and domestic ducks may be bred with a comparatively small mortality, but how great must be the coefficients of regression if their foci of regression are to be placed only as far back, say, as at general populations of jackals and wild ducks.* Apart from cases of atavism, which may be looked upon as improbable variations amply allowed for by theory, we do note, even in dogs, a regression towards a distant ancestry (DARWIN: "Animals and Plants under Domestication," vol. 1, pp. 37, *et seq.*). In these cases, however, change of environment seems in some way more important than the suspension of natural selection. We have, so far, evidence in favour of Mr. GALTON'S view of positions of stability for the focus of regression. It seems, indeed, to be a general opinion among breeders that a character can be fixed, a stock made to breed truer by repeated selection.

Thus DARWIN writes on "Fixedness of Character :" "It is a general belief amongst breeders, that the longer any character has been transmitted by a breed, the more fully it will continue to be transmitted. I do not wish to dispute the truth of

* Professor WEISMANN would place the focus of regression for domestic ducks much further back, presumably in a wingless stage. ("Essays on Heredity," p. 90.)

the proposition that inheritance gains strength simply through long continuance, but I doubt whether it can be proved. In one sense the proposition is little better than a truism; if any character has remained constant during many generations, it will be likely to continue so, if the conditions of life remain the same. So again in improving the breed, if care be taken for a length of time to exclude all inferior individuals, the breed will obviously tend to become truer, as it will not have been crossed during many generations by an inferior animal." ("Animals and Plants under Domestication," vol. 2, p. 37.)

Down to the words "if the conditions of life remain the same," all is consistent with the extreme theory of panmixia, but making a breed *truer* by selection for many generations is only consistent with belief in a progression of the focus of regression, or in a change towards unity in the coefficient of regression with continued selection. The latter alternative would, I think, be quite inconsistent with our whole theory of heredity as applied to a practically stable population. As we cannot mathematically deal with a theory of progression of the focus of regression without some hypothesis of the nature of progression with continued selection, we will assume an extreme case, and suppose the focus to progress very rapidly, *i.e.*, that offspring regress to the mean of the population from which their parents have been immediately selected. This will at least offer some explanation of animals breeding truer with persistent selection, if at the same time it leads to results inconsistent with the extreme theory of panmixia.

(ii.) *Panmixia and Bi-parental Selection.*—Let h_1, s_1 be the paternal, h_2, s_2 the maternal distribution at each selection. Then with assortative mating after p generations, the standard-deviations of the male and female populations will be of the same form as after one generation and be given by the ϵ_1, η_1 of p. 310. Now this result is not like the stable focus of regression out of accord, I think, with experience. It is noteworthy how comparatively little difference there is in the variation constants of the different races of man, although in many cases pretty severe selection may have been supposed to have been in progress for many generations. For example, the mean cephalic index varies from 70 to 83, but the probable deviation from this mean only varies from about 2 to 2·7, so that even very primitive races (where the variation is small and we may suppose the selection has been severe, or the strain is very pure), do not "breed much truer" than highly civilised races with a far less mortality. The difference between the variation of the most and least variable races is probably not more than the β-terms in the values of ϵ_1 and η_1 (p. 310) may be able to account for.

Turning now to the alteration of the male and female means in p-generations of selection, let as before β_2, β_3, β'_2, β'_3, be the regression coefficients and u_n, v_n, the distances from m_2, m_3, of the means of the male and female populations out of which the nth bi-parentage (m_2, m_3, s_2, s_3, ρ) is selected.

Hence: $u_n - (\beta_2 u_n + \beta_3 v_n)$ and $v_n - (\beta'_2 u_n + \beta'_3 v_n)$ are the distances from m_2, m_3

of the means of the male and female populations from which the $n+1^{th}$ bi-parentage is selected.

Thus we have the finite difference equations:

$$u_{n+1} = u_n - (\beta_2 u_n + \beta_3 v_n)$$
$$v_{n+1} = v_n - (\beta'_2 u_n + \beta'_3 v^n)$$

Assume:

$$u_n = A\chi^{n-1}, \quad v_n = \beta\chi^{n-1}.$$

Hence:

$$A(\chi - 1 + \beta_2) = -B\beta_3, \quad B(\chi - 1 + \beta'_3) = -A\beta'_2,$$

or,

$$(\chi - 1)^2 + (\beta_2 + \beta'_3)(\chi - 1) + \beta_2\beta'_3 - \beta_3\beta'_2 = 0,$$

or,

$$\chi_1 = 1 - \gamma_1 \text{ and } \chi_2 = 1 - \gamma_2,$$

where γ_1 and γ_2 have the same values as on p. 311. Thus:

$$u_p = A_1(1 - \gamma_1)^{p-1} + A_2(1 - \gamma_2)^{p-1},$$

$$v_p = A_1\frac{\gamma_1 - \beta_2}{\beta_3}(1 - \gamma_1)^{p-1} + A_2\frac{\gamma_2 - \beta_2}{\beta_3}(1 - \gamma_1)^{p-1};$$

where

$$A_1 = \frac{(\beta_2 - \gamma_2)m_2 + \beta_3 m_3}{\gamma_1 - \gamma_2}$$

$$A_2 = -\frac{(\beta_2 - \gamma_1)m_2 + \beta_3 m_3}{\gamma_1 - \gamma_2}.$$

Now this solution* is the same as that on p. 311, except (i.) that u_p and v_p, unlike μ_p and μ'_p, are measured from the selected means, *i.e.*, the mean heights of the male and female populations are respectively $m_2 - u_p$ and $m_3 - v_p$ after p-generations; (ii.) that in the values u_p and v_p $(1 - \gamma_1)^{p-1}$ and $(1 - \gamma_2)^{p-1}$ replace γ_1^{p-1} and γ_2^{p-1}. We conclude, accordingly, since γ_1, γ_2, and, therefore, $1 - \gamma_1$, $1 - \gamma_2$ are proper fractions, that u_p and v_p grow smaller and smaller, or, if selection be long enough continued, the means of the male and female populations will ultimately pass to the selection means.

Of course, if selection be suspended at the nth generation, regression will take place as on p. 310, but only to the nearest focus of regression, *i.e.*, $m_2 - u_n$, $m_3 - v_n$. Thus the effect of n selections has been to raise the general means permanently by these amounts.

* The uniparental or parthenogenitic results for progression of the focus follow at once by simply putting $\beta_3 = \beta'_2 = \beta'_3 = 0$ in the above formulæ.

(iii.) *Panmixia for Human Stature.*—It is instructive to note the value of these expressions for the case of stature in man. We have at once from the numbers on p. 311, supposing p-selections of male and female populations averaging 4″ and 3″ above the present mean, the following results :

(a.) *Without Assortative Mating.*

$$u_p = 3\cdot9549\,(\cdot2933)^{p-1} + \cdot0451\,(\cdot9581)^{p-1},$$
$$v_p = 3\cdot0538\,(\cdot2933)^{p-1} - \cdot0538\,(\cdot9581)^{p-1}.$$

Thus, in five generations ($p = 5$) $u_5 = \cdot0673$ and $v_5 = -\,\cdot0227$, or the male and female means have been raised 3″·9327 and 3″·0227 respectively. Thus, we see that the males have been raised by selection very near to the selection average, while the females have actually been raised beyond it.* Thus, continued selection would now keep down, and not raise, the female mean, panmixia corresponding to a rise in the mean.

(b.) *With Assortative Mating.*

$$u_p = 3\cdot9893\,(\cdot3560)^{p-1} + \cdot0107\,(\cdot9656)^{p-1}$$
$$v_p' = 3\cdot0136\,(\cdot3560)^{p-1} - \cdot0136\,(\cdot9656)^{p-1}.$$

Thus, in five generations, $u_5 = \cdot0744$ and $v_5 = \cdot0366$, or the male and female means have been raised 3″·9256 and 2″·9634 respectively. The means are accordingly raised less rapidly with this form of sexual relation, the female mean, indeed, having in the five generations not yet overshot the selection mean.

(iv.) *Concluding Remarks on Regression and Fixedness of Character.*—Accordingly on this hypothesis, with the correlation coefficients of inheritance anything like their value in man, five generations of selections of the type required in *both* parents would suffice to establish a breed. This seems more or less consonant with breeders' opinions, which, in part at any rate, may be presumed to represent their experience. If, however, anything like this hypothesis be true, then the suspension of natural selection would not be followed by a rapid regression, or even a slow persistent regression, that would require a reversal of natural selection, *i.e.*, a selection of those previously destroyed and a destruction of those previously selected. On this hypothesis, indeed, it would be probably best to keep the term panmixia for that suspension of assortative mating which we have seen assists, rather than retards, the processes of natural selection.

Several fairly sound reasons could be given why the focus of regression should be taken as the mean of the population from which the parents have been selected, but the sole safe argument appears to be *experiment.*

* This results, of course, from breeding from an average father very much taller relatively than the average mother selected.

The two hypotheses with which we have dealt give practically the two extremes; observation and experiment are perfectly able to determine between them, or to settle whether an intermediate theory is necessary which will give a progression, but a slower progression, to the focus of regression. There are many ways in which analysis can put on the brake, if it be really needful.

At present, all this memoir proposes is to show that such subjects as inheritance, regression, assortative mating and panmixia, are capable of perfectly direct quantitative treatment, and that such treatment, and not somewhat vague discussion of individual instances or of metaphysical possibilities, is what alone can settle the chief problems of evolution. What is wanted is a wide extension of the experimental and statistical work of Mr. FRANCIS GALTON and Professor WELDON. Such numbers as appear in this memoir must be looked upon as illustrative and tentative only. I hope later to publish, for a very limited field, namely, skull measurements in man, a more complete numerical study with mathematical discussion of variation and correlation.

23

Reprinted from pages 209–226 of *Biological Lectures from the Marine Biological Laboratory, Woods Hole, Mass., 1898,* vol. 6, Ginn & Company, Boston, 1899

THE ELIMINATION OF THE UNFIT AS ILLUSTRATED BY THE INTRODUCED SPARROW, *PASSER DOMESTICUS.*

(*A FOURTH CONTRIBUTION TO THE STUDY OF VARIATION.*)

HERMON C. BUMPUS.

We are so in the habit of referring carelessly to the process of natural selection, and of invoking its aid whenever some pet theory seems a little feeble, that we forget we are really using a hypothesis that still remains unproved, and that specific examples of the destruction of animals of known physical disability are very infrequent. Even if the theory of natural selection were as firmly established as Newton's theory of the attraction of gravity, scientific method would still require frequent examination of its claims, and scientific honesty should welcome such examination and insist on its thoroughness.

A possible instance of the operation of natural selection, through the process of the elimination of the unfit, was brought to our notice on February 1 of the present year (1898), when, after an uncommonly severe storm of snow, rain, and sleet, a number of English sparrows were brought to the Anatomical Laboratory of Brown University. Seventy-two of these birds revived; sixty-four perished; and it is the purpose of this lecture to show that the birds which perished, perished not through accident, but because they were physically disqualified, and that the birds which survived, survived because they possessed certain physical characters. These characters enabled them to withstand the intensity of this particular phase of selective elimination, and distinguish them from their more unfortunate companions. It will be convenient for us to arrange our material in the form of tests, as follows.

Test 1: Sex. — It will be noted by reference to the tables that of the *surviving* birds the males are much more numerous than the females. Of the former there are fifty-one (thirty-five adults and sixteen young), while of the latter there are only twenty-one. Among the birds which perished, the females are absolutely and relatively more numerous than they are among the birds which survived, although more than one-half (thirty-six out of sixty-four) of the unfortunate birds are males. Of course it may be that male birds are naturally more abundant than females, but the present question is not one of distribution of sex, but rather of distribution of fitness, and the inference is that the females are less competent to resist severe winter weather than are the males, for, while only 28% of the survivors are females, they constitute 43% of those that perished.

Test 2: Length. — The first column of figures on the several tables gives, in millimeters, the lengths of the birds from the tip of the beak to the tip of the tail. An examination of the averages, printed at the bottom of each column, will prove particularly instructive. It will be noted on Tables I and I[a] that the average length of the adult males which survived (159 mm.) is really less than that of the adult males which perished (**162** mm.).[1] Similar figures, 159 mm. and **162** mm. on Tables II and II[a], indicate the same relative lengths of the young males of the two groups. The average lengths of the females of the two groups, 157 mm. and **158** mm., Tables III and III[a], also indicate an excess in the average length of the birds which perished. The birds which perished, then, males or females, adult or young, are longer than those which endured, and we are justified in concluding that when nature selects, through the agency of winter storms of this particular kind of severity, those sparrows which are relatively short stand a better chance of surviving.

Test 3: Alar Extent. — Averages based on measurements from tip to tip of the extended wings fail to bring out any

[1] The numbers printed in light type, both in the text and in the tables, refer to birds which survived; those printed in heavy type refer to birds which perished.

striking difference between the two classes of birds. Both have an indicated average of 2.45 mm., although, to be more exact, the birds which perished averaged 2.449, while those that survived averaged 2.455, a difference too slight to be of material significance. This similarity of the two groups is not to be wondered at, since it is not to be expected that one eliminative agent will express itself in all possible anatomical features. Were the eliminative agent, for example, a severe northerly wind of protracted duration, the alar extent might then enter in as a factor of considerable selective value, and survivors would then have an alar extent materially different from that of the birds eliminated.

The alar extent of the females, corresponding with their smaller size, is less than that of the males.

Test 4: Weight. — Had I been called upon to express an opinion as to whether heavy or light birds would be more successful in resisting the severity of the February storm, I should have declared unhesitatingly in favor of the heavy birds. An examination of the third column of measurements, however, will show that the birds which survived invariably average less in weight than those which perished, and that among the males this difference amounts to more than a gram; that is, to about one twenty-fifth of the weight. The surviving birds of both sexes had an average weight of 25.2 grams, and those which succumbed had an average weight of 25.8 grams.

It may not be out of place to call attention here to certain objections which may be raised to the method which I have adopted, and to the conclusions thus far derived therefrom. One may claim that the greater relative number of females in the group of birds which perished vitiates the numerical result, since the females are of less stature than the males. But it will be noted that this objection answers itself, for the birds which perished are not shorter, but longer, than those which survived; and again, that the birds which perished, though having a disproportionate number of the lighter sex, nevertheless have an average weight considerably greater than that of the birds which survived. Moreover, comparing, in the two groups, adult males with adult males, young males with young males, and females with females, we find that the differences between the two classes of birds are expressed in these three smaller divisions, and I think we are justified in concluding that the differences are really significant.

The explanation that the birds which lived were those which sought, or at least enjoyed, better shelter cannot be entertained, for the storm was of long duration, and the birds were picked up, not in one locality, but in several localities; and, moreover, it is a fact that the survivors are *structurally different* from those which perished. If to these structural characters one desires to add also the intellectual character that the birds knew enough to go in out of the storm, the difference between the two groups becomes so much the greater.

Test 5: Length of Head. — A comparison of the average lengths of head, from the tip of the beak to the occiput, shows only a similarity between the survivors and those which perished, and indicates that under the present environmental conditions this feature is not sufficiently prominent to be expressed by this method of computation.

Test 6: Length of Humerus. — An examination of the fifth column of figures will show that the length of the arm bones of the birds which perished always averages less than that of the survivors. This difference is most conspicuous in the adult males, where the surviving birds have an average length of humerus of .738 of an inch, considerably more than that of their unfortunate companions, .727.

Here again I wish to emphasize the fact that these differences cannot be merely accidental, because they so often tend in the same direction. If among the survivors it is the proper thing for adult males to have a long humerus, then the young males have a long humerus, and the females follow the prevailing fashion with characteristic servitude. If a short humerus is an index of inferiority, all three groups of eliminated birds (adult males, young males, and females) bear this same mark of inferiority. This fact is the more striking since the averages are established on a relatively small number of birds while usually in the statistical methods of the study of variation an abundance of material is necessary.

Test 7: Length of Femur. — An examination of the general averages on Tables III and III[a] shows that the survivors possess longer thigh bones than do the birds which succumbed. The average length of femur in the former is .716 inch; in the latter .709. This difference in the averages cannot be ascribed to the large number of dead females, since the difference prevails also for both the adult and young males.

Test 8: Length of Tibio-Tarsus. — Measurement of the tibiotarsus yields practically the same comparative data as the measurement of the upper bone of the leg, although in both groups of birds this bone in the females is considerably longer than in the adult males, notwithstanding that the females are smaller. This series of measurements agrees with the sixth, in that the young males have longer legs than the adult males.

Test 9. — Measurements across the skull, from the postorbital bone of one side to the postorbital bone of the other, are given in the eighth column, and are less satisfactory, perhaps, than those of other portions of the skeleton. The breadth of the cranium, as thus indicated, is somewhat less in the females than in the males. The averages denote that the birds which survived had wider heads than those which perished, but these averages are considerably influenced by data furnished by the young males. The irregularities in the subordinate groups induce me to place less confidence in these numerical results than in the results from measurements of other structures.

Test 10: Length of Sternum. — This test differs from other tests in that it relates to measurements in the longitudinal axis of the body. In the males the sternum is long, and in the females it is short. In the birds which survived it has a general average length of .845 inch; in those which perished it has a general average length of only .834.

I think these tests prove that there are fundamental differences between the birds which survived and those which perished. While the former are shorter and weigh less (*i.e.*, are of smaller body), they have longer wing bones, longer legs, longer sternums, and greater brain capacity. These characters are in accordance with our ideas of physical fitness; their defective development is evidently a mark of inferiority, and we are justified in concluding that the birds so handicapped failed to pass one of Nature's rigorous tests and perished.

In an earlier lecture, on the "Variations and Mutations of the Introduced Sparrow," facts were adduced which, it was claimed, were sufficient to show that the English sparrow, since its introduction into this country, has found life so easy that the operation of natural selection has been practically suspended, and that the American type consequently has become degenerate. No active agent had eliminated anomalies, and certain "freaks" had increased in number, until they had become over four times as numerous as in England.

When calling attention to the occurrence of these variations, and to the fact that they were an indication of the absence of an active eliminative factor, I little thought that within a few months I might witness the action of an eliminating factor that would test the structural qualifications of *all* the birds: destroy those which had departed unduly from the ideal type, and thus raise the general standard of excellence.

It will be recalled that, after the storm of February 1, one hundred and thirty-six birds were taken, and that, of these, seventy-two revived, while sixty-four failed to recover. But the fact that the birds which perished had in the *average* longer and larger bodies, and shorter head, wing, and leg bones, does not tell all the story of selective elimination.

Reference to the tables will show, not only that the longest bird perished, but also that the shortest bird perished. The longest bird was No. **33**, the shortest No. **40**. (In these and other cases of extreme departure from the mean, the exponent 1 is placed in the table beside the number of the bird.)

Again, if we examine the columns of figures which indicate the alar extent of the different birds, we find that both the bird with greatest spread of wings, No. **32**, and the one with least spread of wings, No. **52**, perished.

The heaviest bird, No. **23**, weighed 31 grams; it perished. The honors for lightness are evenly divided; No. 53, among the survivors, and No. **60**, among the eliminated, have the same weight, *viz.*, 22.6 grams.

The bird (No. **55**) whose head was longest (measured from the tip of the beak to the occiput) suffered elimination. The extreme variant in the opposite direction (No. 9) survived.

The honors for the longest humerus, .780 mm., are divided, Nos. 6 and **44**. The bird with the shortest humerus, No. **21**, perished.

The longest femur was possessed by bird No. **55**, the shortest by No. 51.

The surviving birds represent both extremes of variation of the tibio-tarsus (Nos. 18 and 41). In respect to all other columns of measurements the survivors possess exclusively never more than one of the *extreme* forms.

Both extremes of variation in width of cranium (Nos. **55** and **52**) are found among the eliminated birds.

The longest sternum is found in one of the surviving birds (No. 15), and it will be remembered that a long sternum was considered a mark of excellence. The shortest sternum (No. **52**) is found among the eliminated birds where the standard is low.

These extremes of variation are represented on Table IV, and by counting the dark numbers we find that eleven extreme positions (maximum or minimum) are occupied exclusively by the birds which perished, whereas the light numbers show that only five extreme positions are occupied by those which survived. In two cases (the minimum weight and the maximum length of humerus) the extreme positions are occupied alike by birds of both groups, and consequently I have left the spaces blank. In three cases two birds of the same group occupy the same extreme position, but the table is designed to indicate only the *extreme positions* and not the number of birds occupying them. The *number* of birds occupying these extreme positions is represented on the previous tables by the exponent 1, and if we count up these exponents, we shall find that among the surviving birds there are nine cases of this extreme type, whereas among the birds which perished there are fourteen cases. These numbers are the more impressive when one considers that, inasmuch as there are seventy-two of the former birds and only sixty-four of the latter, the chances for the occurrence of extreme variation are not equal in the two groups.

The birds which perished are at a decided disadvantage because of their smaller representation, yet there are many more "freaks" among them than among the surviving birds.

If it is thought that the association of the larger number of *extreme* variants with the eliminated birds is merely a matter of accident, we will not stop to argue the matter, but will apply the same test to the birds that remain after these extreme examples have been removed. We find even after the removal of these twenty-three examples, that extreme examples of the second order, indicated by the exponent 2, show the same tendency to occur more frequently among the eliminated birds.

The longest birds now are 166 mm.; the shortest, 153 mm. Of the former, Nos. **22, 24, 28, 32, 35,** perished, and No. 18 survived; of the latter, Nos. **45** and **62** perished, and Nos. 35, 54, and 55 survived.

If we count the times that the exponent 2 occurs in the tables, we shall find that there are ten birds of extreme abnormality of this second grade which survived, while there are twenty of the same grade which perished.

These figures indicate that the amplitude of variation of the surviving birds is less than that of the birds which perished. Were we to attempt the arrangement of the data into curves of distribution, the curve representing the distribution before the storm would be found to have a broad base, whereas the curve representing the distribution after the storm would be found to have a narrow base, for the eliminative process concentrated its energy on the individuals which occupied extreme positions.

Lest there remain some doubt as to the importance of this eliminative process, and of its efficiency in exterminating extreme variants, let us examine our figures again and see whether the group of birds which has already contributed thirty-four of the extremes of variation has still an excess of variability.

If we count up the exponents (3) of this third order of variable individuals, we find that the birds which survive give eleven examples, whereas those which perished give twenty-one. — It appears unnecessary to carry our investigations further along this line, for our results point always in one direction.

Natural selection is most destructive of those birds which have departed most from the ideal type, and its activity raises the general standard of excellence by favoring those birds which approach the structural ideal.

Inasmuch as the variation in structure in the birds which perished tends to centre about certain individuals, as, for example, Nos. 45, 52, and 55, it might be claimed that the accidental presence of a few of these extremely abnormal individuals in this group is what really makes all the difference. Let us see.

There are twenty-three birds among the seventy-two survivors whose measurements bear exponents of extreme variation, and there are twenty-four birds similarly distinguished among the sixty-four which perished. But none of the birds in the first group has more than three exceptional features, whereas several of the birds which perished have a considerably larger number of exceptional features: four, five, and in one case, No. 52, even six.

Of the twenty-three survivors which bear exponents, nineteen have only one exceptional character, and it is not surprising, considering the high standard of excellence possessed by these birds as a whole, that a single unfavorable feature does not prove fatal. There are but ten of the eliminated birds which have only one exceptional character, and the fact that some are burdened with more than one is apparently the reason for their mortality.

In an earlier contribution to the Study of Variation I called attention to a coincidence which may have considerable significance. When specimens of *Necturus* varied in respect to any one feature, there was a tendency for such specimens to present other and not necessarily correlated variations. Stated in another way, instability in respect to any one feature is an index of general organic instability. A similar coincidence of variations occurs among the sparrows.

Of the one hundred and thirty-six birds, five (Nos. 3, 47, 70, 21, 52) had albino feathers. Like other abnormalities endured by the surviving birds, albinism in two out of the three cases is the only affliction. But among those that were eliminated, where albinism twice occurs, it affects in one case a bird marked by four other abnormalities (No. 21), and in the other a bird (No. 52) already cursed by six abnormalities, the most miserable individual in the entire collection.

While we have shown that the birds which perished have certain average structural peculiarities which distinguish them from the survivors, and that the intensity of selective elimination has been felt most by birds of extreme structure, it remains to be shown that a *general instability* of structure is as characteristic of the birds which perished as a *general stability* of structure is characteristic of those which survived.

If we had sufficient data, this fundamental difference in the two groups of birds might be indicated by curves of distribution, one curve narrow and elevated, showing that its components are closely crowded around an ideal mean, the other broad and low, showing that its components are relatively indifferent to any ideal. But in the absence of sufficient data to illustrate the differences in this manner, we can arrive at a numerical result equally instructive by another method.

Having determined the ideal means for the several characters in each group of birds, we can then find the distance that each individual departs from this ideal. By adding these degrees of departure in respect to the several characters, and dividing by the number of individuals, we shall have numbers which represent the *average* departures from the ideal means. These numbers will be large if the members of a group of birds show a general tendency towards *disregard* of the ideals, and they will be small if the birds tend to crowd around the ideals. If all the birds actually attain the ideals, the number will be zero. — This is simply following out the principle that one man at the end of a ten-foot lever can do as much work as ten men at the end of a one-foot lever. A bird removed ten units from the mean exerts the same divergent influence upon its group that ten birds would exercise if removed one unit.

The results of this test, numerically expressed in Table V, are most instructive. In every case but one the numbers indicating the average departure from the ideal mean are smaller for the birds which survived, and thus indicate a general tendency toward conservatism on the part of the survivors. In the single exceptional case the numbers are not very different, 32 and 31. Granting this exception to the uniformity in the figures, it is exceedingly interesting to examine the series. In respect to length, the birds which perished had an average departure from the ideal mean expressed by the number 3.48, while the average departure of the birds which survived was only 2.51, or, expressed in tabular form:

In respect to length,	3.48	is greater than	2.51.
" " " alar extent,	4.60	" " "	4.20.
" " " weight,	12.6	" " "	10.9.
" " " length of head,	5.64	" " "	5.51.
" " " " " humerus,	20.1	" " "	16.0.
" " " " " femur,	20.0	" " "	14.0.
" " " " " tibio-tarsus,	33.8	" " "	29.4.
" " " width of head,	12.	" " "	10., but
" " " length of keel,	31.	" less "	32.

A series of eight consecutive cases like the above, all pointing in the same direction, can hardly be considered accidental.

To summarize:

(1) We have found that there are fundamental differences between the surviving birds and those eliminated, and we conclude that the birds which survived survived because they possessed certain structural characters, and that the birds which perished perished not through accident, but because they did not possess certain structural characters which would have enabled them to withstand the severity of the test imposed by nature; they were eliminated because they were unfit.

(2) The process of selective elimination is most severe with extremely variable individuals, no matter in what direction the variations may occur. It is quite as dangerous to be conspicuously above a certain standard of organic excellence as it is to be conspicuously below the standard. It is the *type* that nature favors.

(3) Disregard of structural qualifications finally produces a throng of degenerates, whose destruction will follow the arrival of adversity.

TABLE I.

Measurements of Thirty-five Males which Survived.

	Total Length.	Alar Extent.	Weight.	Length of Beak and Head.	Length of Humerus.	Length of Femur.	Length of Tibio-tarsus.	Width of Skull.	Length of Keel of Sternum.
1 ♂	154[3]	241	24.5	31.2	.687	.668	1.022[2]	.587	.830
2 ♂	160	252	26.9	30.8	.736	.709	1.180	.602	.841
3 ♂	155	243	26.9	30.6	.733	.704	1.151	.602	.846
4 ♂	154[3]	245	24.3	31.7	.741	.688	1.146	.584	.839
5 ♂	156	247	24.1	31.5	.715	.706	1.129	.575	.821
6 ♂	161	253	26.5	31.8	.780[1]	.743	1.144	.607	.893
7 ♂	157	251	24.6	31.1	.741	.736	1.153	.610	.862
8 ♂	159	247	24.2	31.4	.728	.718	1.126	.609	.793
9 ♂	158	247	23.6	29.8[1]	.703	.673	1.079	.602	.820
10 ♂	158	252	26.2	32.	.749	.739	1.153	.614	.857
11 ♂	160	252	26.2	32.	.741	.723	1.129	.624	.892
12 ♂	162	253	24.8	32.3	.766	.752	1.134	.633	.923[2]
13 ♂	161	243	25.4	31.8	.721	.722	1.126	.597	.891
14 ♂	160	250	23.7	29.8[1]	.730	.703	1.103	.590	.820
15 ♂	159	247	25.7	31.4	.729	.717	1.141	.592	.927[1]
16 ♂	158	253	25.7	31.9	.743	.699	1.150	.600	.360
17 ♂	159	247	26.5	31.6	.733	.714	1.155	.611	.923[2]
18 ♂	166[2]	253	26.7	32.5	.767	.765[2]	1.230[1]	.600	.878
19 ♂	159	247	23.9	31.4	.752	.723	1.113	.602	.825
20 ♂	160	248	24.7	31.3	.752	.737	1.176	.603	.803
21 ♂	161	252	28.	31.8	.770[2]	.731	1.190	.590	.885
22 ♂	163	251	27.9	31.9	.769[8]	.745	1.168	.622	.860
23 ♂	156	242	25.9	32.	.723	.711	1.116	.609	.886
24 ♂	165[8]	251	25.7	32.2	.751	.742	1.161	.613	.865
25 ♂	160	247	26.6	32.4	.728	.707	1.108	.590	.836
26 ♂	158	244	23.2[8]	31.6	.730	.713	1.142	.585	.888
27 ♂	160	242	25.7	31.6	.709	.705	1.124	.620	.788
28 ♂	157	245	26.3	32.2	.741	.726	1.143	.595	.850
29 ♂	159	244	24.3	31.5	.723	.698	1.107	.615	.847
30 ♂	160	253	26.7	32.1	.739	.714	1.117	.592	.864
31 ♂	158	245	24.9	31.4	.726	.703	1.119	.580	.854
32 ♂	161	247	23.8	31.4	.735	.694	1.101	.602	.789
33 ♂	160	247	25.6	32.3	.756	.745	1.135	.607	.902
34 ♂	160	247	27.	32.	.755	.736	1.174	.631	.873
35 ♂	153[2]	241	24.7	32.2	.728	.680	1.092	.592	.884
Average . .	159	247	25.4	31.6	.738	.716	1.135	.602	.857

TABLE I[a].

Measurements of Twenty-four Adult Males which Perished.

	TOTAL LENGTH.	ALAR EXTENT.	WEIGHT.	LENGTH OF BEAK AND HEAD.	LENGTH OF HUMERUS.	LENGTH OF FEMUR.	LENGTH OF TIBIO-TARSUS.	WIDTH OF SKULL.	LENGTH OF KEEL OF STERNUM.
1 ♂	165^{3}	249	26.5	31.	.738	.704	1.095	.606	.847
2 ♂	160	245	26.1	32.	.736	.709	1.109	.611	.842
3 ♂	161	249	25.6	32.3	.743	.718	1.128	.602	.828
4 ♂	162	246	25.9	32.3	.738	.709	1.135	.607	.869
5 ♂	163	250	25.5	32.5	.752	.731	1.197	.623	.888
6 ♂	162	247	27.6	31.8	.731	.719	1.113	.597	.869
7 ♂	163	246	25.8	31.4	.689	$.662^{3}$	1.073	.604	.836
8 ♂	161	246	24.9	30.5	.739	.726	1.138	.580	.803
9 ♂	160	242	26.	31.	.745	.713	1.105	.600	.803
10 ♂	162	246	26.5	31.5	.720	.696	1.092	.606	.809
11 ♂	160	249	26.	31.4	.726	.689	1.097	.602	.850
12 ♂	161	250	27.1	31.6	.737	.711	1.120	.631	.852
13 ♂	162	248	25.1	31.9	.744	.722	1.154	.591	.839
14 ♂	165^{3}	252	26.	32.3	.726	.710	1.145	.609	.887
15 ♂	161	243	25.6	32.5	.709	.707	1.122	.607	.832
16 ♂	161	244	25.	31.3	.702	.685	1.082	.595	.874
17 ♂	162	248	24.6	31.	.713	.700	1.086	.590	.837
18 ♂	164	244	25.	31.2	.703	.690	1.074	.608	.795
19 ♂	158	247	26.	32.	.729	.710	1.145	.607	.803
20 ♂	162	253	28.3	31.8	.752	.718	1.152	.600	.857
21 ♂	156	239	24.6	30.5	$.659^{1}$	$.658^{2}$	1.042^{3}	$.570^{3}$	.810
22 ♂	166	251	27.5	31.5	.720	.691	1.118	.612	.847
23 ♂	165^{3}	253	31. 1	32.4	.765	.750	1.183	.613	.905
24 ♂	166^{2}	250	28.3	32.4	.754	.718	1.179	.607	$.916^{3}$
Average . .	162	247	26.2	31.6	.727	.706	1.120	.603	.845

TABLE II.

Measurements of Sixteen Young Males which Survived.

	Total Length.	Alar Extent.	Weight.	Length of Beak and Head.	Length of Humerus.	Length of Femur.	Length of Tibio-tarsus.	Width of Skull.	Length of Keel of Sternum.
36 juv. ♂	156	246	24.6	32.	.741	.735	1.167	.592	.849
37 juv. ♂	156	245	25.5	32.1	.761	.717	1.147	.620	.816
38 juv. ♂	163	248	24.8	32.2	.742	.733	1.165	.606	.854
39 juv. ♂	163	248	26.3	33.[3]	.736	.704	1.148	.609	.839
40 juv. ♂	160	250	24.4	31.5	.746	.715	1.173	.604	.893
41 juv. ♂	156	237	23.3	30.6	.692	.664	1.011[1]	.588	.774
42 juv. ♂	162	253	26.7	32.	.759	.734	1.197	.630	.878
43 juv. ♂	163	254[3]	26.4	32.	.766	.750	1.165	.605	.886
44 juv. ♂	164	251	26.9	32.	.755	.742	1.171	.620	.886
45 juv. ♂	163	244	24.3	31.3	.718	.680	1.082	.610	.892
46 juv. ♂	160	247	27.	31.5	.764	.732	1.177	.617	.846
47 juv. ♂	160	250	26.8	32.5	.764	.729	1.123	.635[2]	.842
48 juv. ♂	158	247	24.9	32.4	.745	.724	1.139	.588	.865
49 juv. ♂	158	249	26.1	32.2	.742	.736	1.148	.602	.817
50 juv. ♂	158	243	26.6	32.4	.747	.711	1.163	.612	.891
51 juv. ♂	155	237	23.3	30.2[3]	.685	.653[1]	1.011[1]	.587	.794
Average . .	159	246	25.4	31.8	.741	.716	1.136	.607	.851

TABLE II[a].

Measurements of Twelve Young Males which Perished.

	Total Length.	Alar Extent.	Weight.	Length of Beak and Head.	Length of Humerus.	Length of Femur.	Length of Tibio-tarsus.	Width of Skull.	Length of Keel of Sternum.
25 juv. ♂	160	249	24.2	30.4	.740	.717	1.130	.620	.840
26 juv. ♂	156	236	26.8	$30.^{2}$	.690	.671	1.067	$.563^{2}$	.832
27 juv. ♂	158	240	23.5	31.	.715	.702	1.113	.595	.805
28 juv. ♂	166^{2}	245	26.9	31.7	.715	.695	1.107	.601	.847
29 juv. ♂	165^{3}	255^{2}	28.6	31.5	.766	.744	1.175	.613	.854
30 juv. ♂	157	238	24.7	31.2	$.680^{3}$	.677	1.156	.599	.769
31 juv. ♂	164	250	27.3	31.8	.764	.726	1.171	.588	.860
32 juv. ♂	166^{2}	256^{1}	25.7	31.7	.752	.751	1.187	.595	.858
33 juv. ♂	167^{1}	255^{2}	$29.^{3}$	32.2	.765	.745	1.197	$.638^{2}$	.855
34 juv. ♂	161	246	25.	31.5	.739	.707	1.123	.587	.850
35 juv. ♂	166^{2}	254^{3}	27.5	31.4	.760	.742	1.124	.604	.914
36 juv. ♂	161	251	26.	31.5	.731	.707	1.122	.589	.828
Average . .	162	248	26.2	31.5	.734	.715	1.141	.599	.842

Table III.

Measurements of Twenty-one Adult and Young Females which Survived.

	Total Length.	Alar Extent.	Weight.	Length of Beak and Head.	Length of Humerus.	Length of Femur.	Length of Tibio-tarsus.	Width of Skull.	Length of Keel of Sternum.
52 ♀	156	245	25.3	31.6	.729	.710	1.152	.620	.809
53 ♀	154^{8}	240	22.6^{1}	30.4	.705	.686	1.103	.584	.770
54 ♀	153^{2}	240	25.1	31.	.724	.713	1.123	.585	.812
55 ♀	153^{2}	236	23.2^{3}	30.9	.698	.678	1.132	.596	795
56 ♀	155	243	24.4	31.5	.734	.736	1.170	.596	801
57 ♀	163	247	25.1	32.	.748	.734	1.166	.602	.821
58 ♀	157	238	24.6	30.9	.726	.727	1.175	.588	.797
59 ♀	155	239	24.	32.8	.732	.742	1.175	.601	.835
60 ♀	164	248	24.2	32.7	.752	.752	1.201	.604	.830
61 ♀	158	238	24.9	31.	.741	.689	1.091	.592	.866
62 ♀	158	240	24.1	31.3	.733	.706	1.107	.591	.867
63 ♀	160	244	24.	31.1	.731	.730	1.152	.589	.808
64 ♀	161	246	26.	32.3	.758	.732	1.154	.623	.859
65 ♀	157	245	24.9	32.	.752	.740	1.186	.593	.787
66 ♀	157	235	25.5	31.5	.712	.704	1.132	.611	.781
67 ♀	156	237	23.4	30.9	.708	.691	1.123	.613	.798
68 ♀	158	244	25.9	31.4	.729	.705	1.146	.597	.851
69 ♀	153^{2}	238	24.2	30.5	.715	.707	1.116	.595	.821
70 ♀	155	236	24.2	30.3	.727	.705	1.120	.585	.790
71 ♀	163	246	27.4	32.5	.732	.711	1.163	.630	.862
72 ♀	159	236	24.	31.5	.709	.713	1.129	.607	.845
Average . .	157	241	24.6	31.4	.728	.714	1.143	.600	.819
General average for 72 birds . . .	158	245	25.2	31.6	.736	.716	1.138	.603	.845

TABLE III^{a}.

Measurements of Twenty-eight Adult and Young Females which Perished.

	TOTAL LENGTH.	ALAR EXTENT	WEIGHT.	LENGTH OF BEAK AND HEAD.	LENGTH OF HUMERUS.	LENGTH OF FEMUR.	LENGTH OF TIBIO-TARSUS.	WIDTH OF SKULL.	LENGTH OF KEEL OF STERNUM.
37 ♀	155	240	26.3	31.4	.709	.710	1.125	.614	.815
38 ♀	156	240	25.8	31.5	.715	.678	1.127	.597	.812
39 ♀	160	242	26.	32.6	.740	.732	1.157	.597	.854
40 ♀	152^{1}	232^{3}	23.2^{3}	30.3	.676^{2}	.683	1.048	.590	.780
41 ♀	160	250	26.5	31.7	.741	.731	1.187	.615	.886
42 ♀	155	237	24.2	31.	.727	.723	1.118	.610	.787
43 ♀	157	245	26.9	32.2	.766	.751	1.227^{2}	.620	.841
44 ♀	165^{3}	245	27.7	33.1^{2}	.780^{1}	.757^{3}	1.195	.633	.895
45 ♀	153^{2}	231^{2}	23.9	30.1	.680^{3}	.662^{3}	1.042^{3}	.592	.781
46 ♀	162	239	26.1	30.3	.709	.685	1.092	.587	.911
47 ♀	162	243	24.6	31.6	.741	.729	1.162	.605	.840
48 ♀	159	245	23.6	31.8	.727	.700	1.129	.610	.855
49 ♀	159	247	26.	30.9	.711	.666	1.098	.580	.749^{2}
50 ♀	155	243	25.	30.9	.730	.711	1.127	.598	.839
51 ♀	162	252	24.8	31.9	.752	.738	1.180	.615	.875
52 ♀	152^{1}	230^{1}	22.8^{2}	30.4	.682	.664	1.042^{3}	.551^{1}	.734^{1}
53 ♀	159	242	24.8	30.8	.717	.667	1.090	.575	.809
54 ♀	155	238	24.6	31.2	.706	.702	1.102	.588	.758^{3}
55 ♀	163	249	30.5^{2}	33.4^{1}	.767	.767^{1}	1.207^{3}	.640^{1}	.896
56 ♀	163	242	24.8	31.	.713	.713	1.128	.607	.813
57 ♀	156	237	23.9	31.7	.718	.716	1.090	.611	.800
58 ♀	159	238	24.7	31.5	.726	.701	1.145	.600	.800
59 ♀	161	245	26.9	32.1	.751	.704	1.142	.607	.819
60 ♀	155	235	22.6^{1}	30.7	.695	.692	1.119	.584	.771
61 ♀	162	247	26.1	31.9	.751	.735	1.157	.618	.802
62 ♀	153^{2}	237	24.8	30.6	.732	.718	1.172	.594	.802
63 ♀	162	245	26.2	32.5	.728	.731	1.102	.614	.832
64 ♀	164	248	26.1	32.3	.739	.707	1.159	.592	.823
Average . .	158	241	25.3	31.4	.726	.709	1.131	.601	.820
General average for 64 birds . . .	160	245	25.8	31.5	.728	.709	1.128	.601	.834

TABLE IV.

The Maximum and Minimum Measurements.

	Total Length.	Alar Extent.	Weight.	Length of Beak and Head.	Length of Humerus.	Length of Femur.	Length of Tibio-tarsus.	Width of Skull.	Length of Keel of Sternum.
Maximum . .	167	256	31	33.4	(*d*)	.767	1.230	.640	927
Minimum . .	(*a*) 152	230	(*b*)	(*c*) 29.8	.659	.653	(*c*) 1.011	.551	.734

(*a*) The minimum length (152 mm.) occurs twice among the birds which perished: Nos. 40 and 52.
(*b*) The minimum weight (22.6 grams) occurs in each group: Nos. 53 and 60, and therefore is not entered.
(*c*) The minimum length of head (29.8 mm.) and of tibio-tarsus (1.011 inch) occurs twice among the surviving birds: Nos. 9 and 14, 41 and 51.
(*d*) The maximum length of humerus (.780 inch) occurs in each group: Nos. 6 and 44, and therefore is not entered.

TABLE V.

Average Departures from Ideal Mean.

	Total Length.	Alar Extent.	Weight.	Length of Beak and Head.	Length of Humerus.	Length of Femur.	Length of Tibio-tarsus.	Width of Skull.	Length of Keel of Sternum.
Seventy-two which survived	2.51	4.20	10.9	5.51	16.	14.	29.4	10.	32.
Sixty four which perished	3.48	4.60	12.6	5.64	20.1	20.	33.8	12.	31.

EPILOGUE

While Charles Darwin had convinced the majority of scholars by 1870 that evolution had occurred, most of them considered natural selection to be a relatively unimportant cause of evolution. Some scholars championed vitalistic theories of evolution (Cope, 1887, 1896; Eimer, 1890, 1898; Mivart, 1871) while many advocated Lamarckian theories and contended that adaptation was produced by the inheritance of characters acquired by the use and disuse of organs. Those who championed nonselectionist theories of evolution did so primarily because they believed that the natural processes producing individual variations rather than those generating selection are the ones that determine the direction of evolution. Many neo-Lamarckians claimed that an animal could modify its structure by its conscious efforts and that such acquired modifications were genetically inherited. In contrast two psychologists, James M. Baldwin (1896) and C. Lloyd Morgan (1896), and a paleontologist, Henry F. Osborn (1896), each independently developed the same explicit theory (now called the "Baldwin Effect") to explain how an individual organism's behavior could affect the selective forces operating on it; that is, how an individual organism's accomodation to an environmental situation may "under the influence of natural selection be reinforced or replaced by similar hereditary characteristics" (Simpson, 1953*a*, p. 110).

The rediscovery of the Mendelian principles of heredity in 1900 brought about an almost total eclipse of Lamarckian theories of evolution by 1930. Initially, most of the Mendelians also opposed the Darwinian theory of gradual evolution by the natural selection of small variations. They embraced the mutationist theory of speciation championed by Hugo DeVries (1901–1903, 1904). The Mendelians abandoned their initial objections to selection as a creative process directing evolution when, "as the result of their own research, they came to realize that Mendelian heredity, far from being antagonistic to Darwinian selection, was in reality, complementary to it" (Provine, 1971; see also Bajema, 1982, pp. 222–326).

Mathematical models that described the effects of selection, mating systems, mutation, genetic drift, and migration on the frequency and distribution of genes in populations were developed between 1918 and 1932 by Chetverikov (1926, 1927), Fisher (1918, 1930), Haldane (1924–1932, 1932), and Wright (1921, 1931, 1932). These population genetics models enabled biologists to quantitatively compare the relative importance of selection, mutation, and genetic drift in bringing about genetic change in populations. The mathematical modeling of evolutionary processes in populations by Fisher, Haldane, and Wright (1) helped convince scientists that very small selection pressures were capable of genetically changing populations rapidly enough to have caused the evolutionary changes observed in the fossil record, and (2) eliminated many competing theories by showing that Lamarckian, orthogenetic, and other hypothesized factors were not logically essential for evolution to occur (Provine, 1978).

The synthetic theory of evolution that developed during the 1930s, 1940s, and 1950s provided an explanation of phylogeny, ongoing evolution, and speciation that incorporated scientific knowledge concerning wild populations gathered by naturalists, and the experimental manipulation of populations of laboratory, domesticated, and wild organisms. The theory also was based on the population genetics models of Mendelian populations constructed by Fisher, Haldane, and Wright (Provine, 1978; Mayr and Provine, 1980). The development of the synthetic theory of evolution was brought about by many scientists including Dobzhansky (1937), Huxley (1942), Mayr (1942), Simpson, (1944, 1953*b*), and Stebbins (1950).

The scientific literature on natural selection becomes so voluminous after 1900 that it is impossible to reprint in one volume all the benchmark quality papers that have advanced our scientific understanding of the theory and measurement of natural selection. A companion Benchmark volume on the twentieth-century theory and measurement of natural selection will include papers that deal primarily with the ecological causes of natural selection. Benchmark volumes by Jameson (1977*b*), and Weiss and Ballonoff (1975) reprint numerous papers that provide insight into the operation of natural selection, while the Benchmark volume by Bajema (1976) contains papers that have attempted to apply selection theory to the problems of eugenics—the future genetic evolution of human populations. Benchmark volumes by Jameson (1977*a*) and Levin (1979) reprint papers providing insight into the role of natural selection in speciation. Papers on the role that sampling error (genetic drift) plays in evolution can be found in Li (1977) while papers on the nature of mutations have been reprinted by Drake and Koch (1976).

This volume includes reprints of many of the classic papers on natural selection that were published prior to 1900. Other Benchmark volumes on artificial selection, sexual selection, sexual selection and the development of sociobiology, human evolution by selection, cultural evolution by selection, and the domestication of plants and animals by selection have been published or are in preparation. A Benchmark volume on the evolution of mimicry as the result of selection generated by predator-prey ecological interactions is also in preparation.

Our scientific understanding of adaptive evolution by selection has undergone a radical change since Charles Darwin first championed his theory in the *Origin of Species.* Selection has been transformed by Hamilton (1964*a*, 1964*b*), Trivers (1971, 1972, 1974), Maynard Smith (1972) and others from a mere survival principle ("survival of the fittest") into a reproduction maximizing principle (inclusive fitness) with tremendous implications for the study of social behavior of human beings as well as of other species.

REFERENCES

Bajema, C. J., ed., 1976, *Eugenics: Then and Now,* Benchmark Papers in Genetics, vol. 5, Dowden, Hutchinson & Ross, Stroudsburg, Pa., 382p.

Bajema, C. J., ed., 1982, *Artificial Selection and the Development of Evolutionary Theory,* Benchmark Papers in Systematic and Evolutionary Biology, vol. 4, Hutchinson Ross, Stroudsburg, Pa., 384p.

Baldwin, J., 1896, A New Factor in Evolution, *Am. Nat.* **30:**441–451 536–553.

Chetverikov, S. S., 1926, On Certain Aspects of the Evolutionary Process from the Standpoint of Modern Genetics (in Russian), *Zh. Eksp. Biol.* **A2:**3–54. (An English translation appears in *Am. Philos. Soc. Proc.* **105:**167–195, 1961, and is reprinted as Paper 18 in Jameson, 1977*b*.)

Chetverikov, S. S., 1927, On the Genetic Constitution of Wild Populations (in Russian), *Int. Congr. Genetics, 1927, 5th, Proc.,* pp. 1499–1500. (An English translation appears as Paper 19 in Jameson, 1977*b*.)

Cope, E. D., 1887, *The Origin of the Fittest: Essays on Evolution,* Appleton, New York, 467p.

Cope, E. D., 1896, *The Primary Factors of Evolution,* Open Court, Chicago, 547p.

de Vries, H., 1901–1903, *Die Mutationstheorie. Versuche und Beobachtungen Uber die Enstehung der Arten im Pflanzenreich,* 2 vols., Veit & Co., Leipzig, 648p., 752p. (English translation by J. Farmer and A. Darbishire, 1910, Open Court, Chicago.)

de Vries, H., 1904, *Species and Varieties: Their Origin by Mutation,* Open Court, Chicago, 847p.

Drake, J. W., and R. E. Koch, eds., 1976, *Mutagenesis,* Benchmark Papers in Genetics, vol. 4, Dowden, Hutchinson & Ross, Stroudsburg, Pa., 363p.

Eimer, G. H. T., 1890, *Organic Evolution as the Result of the Inheritance of Acquired Characters According to the Laws of Organic Growth,* Macmillan, New York, 435p.

Eimer, G. H. T., 1898, *On Orthogenesis and the Impotence of Natural Selection in Species-Formation,* Open Court, Chicago, 56p.

Fisher, R. A., 1918, The Correlation Between Relatives on the Supposition of Mendelian Inheritance, *Roy. Soc. Edinburgh Trans.* **52:**399–433.
Fisher, R. A., 1930, *The Genetical Theory of Natural Selection,* Oxford University Press, Oxford, 272p.
Haldane, J. B. S., 1924–1932, A Mathematical Theory of Natural and Artificial Selection, Part I, *Cambridge Philos. Soc. Proc.* **23:**19–41; Part II, **23:**158–163; Part III, **23:**363–372; Part IV, **23:**607–615; Part V, **23:**838–844; Part VI, **26:**220–230; Part VII, **27:**131–136; Part VIII, **27:**137–142; Part IX, **28:**244–248. (Part VI appears as Paper 12 in Jameson, 1977*a* and Parts I, II, III, VII, VIII, and IX appear as Papers 15, 16, 17, 21, 22, and 23 in Jameson, 1977*b*.)
Haldane, J. B. S., 1932, *The Causes of Evolution,* Longmans, Green & Co., London, 235p.
Hamilton, W. D., 1964*a*, The Genetical Theory of Social Behaviour. I., *J. Theor. Biol.* **7:**1–16.
Hamilton, W. D., 1964*b*, The Genetical Theory of Social Behaviour. II., *J. Theor. Biol.* **7:**17–52.
Huxley, J., 1942, *Evolution: The Modern Synthesis,* Harper, New York, 645p.
Jameson, D. L., ed., 1977*a*, *Genetics of Speciation,* Benchmark Papers in Genetics, vol. 7, Dowden, Hutchinson & Ross, Stroudsburg, Pa., 336p.
Jameson, D. L., ed., 1977*b*, *Evolutionary Genetics,* Benchmark Papers in Genetics, vol. 8, Dowden, Hutchinson & Ross, Stroudsburg, Pa., 332p.
Levin, D., ed., 1979, *Hybridization: An Evolutionary Perspective,* Benchmark Papers in Genetics, vol. 11, Dowden, Hutchinson & Ross, Stroudsburg, Pa., 321p.
Li, W. H., ed., 1977, *Stochastic Models in Population Genetics,* Benchmark Papers in Genetics, vol. 7, Dowden, Hutchinson & Ross, Stroudsburg, Pa., 471p.
Maynard Smith, J., 1972, Game Theory and the Evolution of Fighting, in *On Evolution,* Edinburgh University Press, Edinburgh, pp. 8–28.
Mayr, E., 1942, *Systematics and the Origin of Species,* Columbia University Press, New York, 334p.
Mayr, E., and W. B. Provine, eds., 1980, *The Evolutionary Synthesis: Perspectives on the Unification of Biology,* Harvard University Press, Cambridge, Mass., 487p.
Mivart, St. G., 1871, *On the Genesis of Species,* Appleton, New York, 314p.
Morgan, C. L., 1896, *Habit and Instinct,* Arnold, London, 351p.
Osborn, H. F., 1896, A Mode of Evolution Requiring Neither Natural Selection nor the Inheritance of Acquired Characteristics, *N.Y. Acad. Sci. Trans.* **15:**141–142.
Provine, W. B., 1971, *The Origins of Theoretical Population Genetics,* University of Chicago Press, Chicago, 201p.
Provine, W. B., 1978, The Role of Mathematical Population Geneticists in the Evolutionary Synthesis of the 1930s and 1940s, *Stud. Hist. Biol.* **2:**167–192.
Simpson, G. G., 1944, *Tempo and Mode in Evolution,* Columbia University Press, New York, 237p.
Simpson, G. G., 1953*a*, The Baldwin Effect, *Evolution* **7:**110–117.
Simpson, G. G., 1953*b*, *The Major Features of Evolution,* Columbia University Press, New York, 434p.
Stebbins, G. L., 1950, *Variation and Evolution in Plants,* Columbia University Press, New York, 643p.

Trivers, R. L., 1971, The Evolution of Reciprocal Altruism, *Quart. Rev. Biol.* **46:**35-57.
Trivers, R. L., 1972, Parental Investment and Sexual Selection, in *Sexual Selection and the Descent of Man, 1871-1971,* B. Campbell, ed., Aldine, Chicago, pp. 136-179.
Trivers, R. L., 1974, Parent-Offspring Conflict, *Am. Zool.* **14:**249-264.
Weiss, K. M., and P. A. Ballonoff, eds., 1975, *Demographic Genetics,* Benchmark Papers in Genetics, vol. 3, Dowden, Hutchinson & Ross, Stroudsburg, Pa., 414p.
Wright, S., 1921, Systems of Mating, *Genetics* **6:**111-178.
Wright, S., 1931, Evolution in Mendelian Populations, *Genetics* **16:**97-159.
Wright, S., 1932, The Roles of Mutation, Inbreeding, Crossbreeding and Selection in Evolution, *Int. Congr. Genetics, 1932, 6th, Proc.* **1:**356-366.

AUTHOR CITATION INDEX

SUBJECT INDEX

About the Editor

CARL JAY BAJEMA is professor of biology in the College of Arts and Sciences at Grand Valley State Colleges, Allendale, Michigan. He teaches courses in human ecology, human genetics, human sexuality, evolutionary biology, and sociobiology in the Department of Biology, and the Darwinian revolution in the History of Science program. Professor Bajema earned the Ph.D. in zoology from Michigan State University in 1963.

Professor Bajema has served as Senior Population Council Fellow in Demography and Population Genetics at the University of Chicago (1966-1967); as Research Associate in Population Studies at Harvard University (1967-1972); as Research Associate in Biology at the University of California, Santa Barbara (1974); and most recently as Visiting Professor of Anthropology at Harvard University (1974-1975).

Dr. Bajema's research interests have concentrated on the estimation of the direction and intensity of natural selection with respect to human behavior. Professor Bajema is editor of *Natural Selection in Human Populations* (Wiley, 1971) and coauthor with Garrett Hardin of the third edition of the introductory college biology textbook *Biology: Its Principles and Implications* (Freeman, 1978). He has edited a Benchmark Papers in Genetics volume, *Eugenics Then and Now* (Dowden, Hutchinson & Ross, 1976) and a Benchmark Papers in Systematic and Evolutionary Biology volume, *Artificial Selection and the Development of Evolutionary Theory* (Hutchinson Ross, 1982), in addition to this volume. Professor Bajema is editing additional Benchmark volumes on selection theory. He is series editor for the Benchmark Papers in Systematic and Evolutionary Biology series being published by Hutchinson Ross Publishing Company.

Professor Bajema enjoys hiking, bicycling, and cross-country skiing. He resides in Grand Rapids, Michigan.